Energy of Matter

Joseph A. Angelo, Jr.

Facts On File
An Infobase Learning Company

This book is dedicated to my sons, Joseph and James.

ENERGY OF MATTER

Copyright © 2011 by Joseph A. Angelo, Jr.

Facts On File, Inc.
An imprint of Infobase Learning
132 West 31st Street
New York NY 10001

Library of Congress Cataloging-in-Publication Data
Angelo, Joseph A.
   Energy of matter / Joseph A. Angelo, Jr.
      p. cm.—(States of matter)
   Includes bibliographical references and index.
   ISBN 978-0-8160-7605-5
   1. Matter—Properties. 2. Force and energy. I. Title.
   QC173.39.A54 2011
   531'.6—dc22   2010044118

Facts On File books are available at special discounts when purchased in bulk quantities for businesses, associations, institutions, or sales promotions. Please call our Special Sales Department in New York at (212) 967-8800 or (800) 322-8755.

You can find Facts On File on the World Wide Web at http://www.infobaselearning.com

Text design by Annie O'Donnell
Composition by Hermitage Publishing Services
Illustrations by Sholto Ainslie
Photo research by the author
Cover printed by Yurchak Printing, Inc., Landisville, Pa.
Book printed and bound by Yurchak Printing, Inc., Landisville, Pa.
Date printed: August 2011
Printed in the United States of America

10 9 8 7 6 5 4 3 2 1

This book is printed on acid-free paper.

Contents

Preface

The unleashed power of the atom has changed everything save our modes of thinking.

—Albert Einstein

Humankind's global civilization relies upon a family of advanced technologies that allow people to perform clever manipulations of matter and energy in a variety of interesting ways. Contemporary matter manipulations hold out the promise of a golden era for humankind—an era in which most people are free from the threat of such natural perils as thirst, starvation, and disease. But matter manipulations, if performed unwisely or improperly on a large scale, can also have an apocalyptic impact. History is filled with stories of ancient societies that collapsed because local material resources were overexploited or unwisely used. In the extreme, any similar follies by people on a global scale during this century could imperil not only the human species but all life on Earth.

Despite the importance of intelligent stewardship of Earth's resources, many people lack sufficient appreciation for how matter influences their daily lives. The overarching goal of States of Matter is to explain the important role matter plays throughout the entire domain of nature—both here on Earth and everywhere in the universe. The comprehensive multivolume set is designed to raise and answer intriguing questions and to help readers understand matter in all its interesting states and forms—from common to exotic, from abundant to scarce, from here on Earth to the fringes of the observable universe.

The subject of matter is filled with intriguing mysteries and paradoxes. Take two highly flammable gases, hydrogen (H_2) and oxygen (O_2), carefully combine them, add a spark, and suddenly an exothermic reaction takes place yielding not only energy but also an interesting new substance called water (H_2O). Water is an excellent substance to quench a fire, but it is also an incredibly intriguing material that is necessary for all life here on Earth—and probably elsewhere in the universe.

Matter is all around us and involves everything tangible a person sees, feels, and touches. The flow of water throughout Earth's biosphere, the air people breathe, and the ground they stand on are examples of the most commonly encountered states of matter. This daily personal encounter with matter in its liquid, gaseous, and solid states has intrigued human beings from the dawn of history. One early line of inquiry concerning the science of matter (that is, *matter science*) resulted in the classic earth, air, water, and fire elemental philosophy of the ancient Greeks. This early theory of matter trickled down through history and essentially ruled Western thought until the Scientific Revolution.

It was not until the late 16th century and the start of the Scientific Revolution that the true nature of matter and its relationship with energy began to emerge. People started to quantify the properties of matter and to discover a series of interesting relationships through carefully performed and well-documented experiments. Speculation, philosophical conjecture, and alchemy gave way to the scientific method, with its organized investigation of the material world and natural phenomena.

Collectively, the story of this magnificent intellectual unfolding represents one of the great cultural legacies in human history—comparable to the control of fire and the invention of the alphabet. The intellectual curiosity and hard work of the early scientists throughout the Scientific Revolution set the human race on a trajectory of discovery, a trajectory that not only enabled today's global civilization but also opened up the entire universe to understanding and exploration.

In a curious historical paradox, most early peoples, including the ancient Greeks, knew a number of fundamental facts about matter (in its solid, liquid, and gaseous states), but these same peoples generally made surprisingly little scientific progress toward unraveling matter's inner mysteries. The art of metallurgy, for example, was developed some 4,000 to 5,000 years ago on an essentially trial-and-error basis, thrusting early civilizations around the Mediterranean Sea into first the Bronze Age and later the Iron Age. Better weapons (such as metal swords and shields) were the primary social catalyst for technical progress, yet the periodic table of chemical elements (of which metals represent the majority of entries) was not envisioned until the 19th century.

Starting in the late 16th century, inquisitive individuals, such as the Italian scientist Galileo Galilei, performed careful observations and measurements to support more organized inquiries into the workings of the natural world. As a consequence of these observations and experiments, the nature of matter became better understood and better quantified. Scientists introduced the concepts of density, pressure, and temperature in their efforts to more consistently describe matter on a large (or macroscopic) scale. As instruments improved, scientists were able to make even better measurements, and soon matter became more clearly understood on both a macroscopic and microscopic scale. Starting in the 20th century, scientists began to observe and measure the long-hidden inner nature of matter on the atomic and subatomic scales.

Actually, intellectual inquiry into the microscopic nature of matter has its roots in ancient Greece. Not all ancient Greek philosophers were content with the prevailing earth-air-water-fire model of matter. About 450 B.C.E., a Greek philosopher named Leucippus and his more well-known student Democritus introduced the notion that all matter is actually composed of tiny solid particles, which are *atomos* (ατομος), or indivisible. Unfortunately, this brilliant insight into the natural order of things lay essentially unnoticed for centuries. In the early 1800s, a British schoolteacher named John Dalton began tinkering with mixtures of gases and made the daring assumption that a chemical element consisted of identical indestructible atoms. His efforts revived atomism. Several years later, the Italian scientist Amedeo Avogadro announced a remarkable hypothesis, a bold postulation that paved the way for the atomic theory of chemistry. Although this hypothesis was not widely accepted until the second half of the 19th century, it helped set the stage for the spectacular revolution in matter science that started as the 19th century rolled into the 20th.

What lay ahead was not just the development of an atomistic kinetic theory of matter, but the experimental discovery of electrons, radioactivity, the nuclear atom, protons, neutrons, and quarks. Not to be outdone by the nuclear scientists, who explored nature on the minutest scale, astrophysicists began describing exotic states of matter on the grandest of cosmic scales. The notion of degenerate matter appeared as well as the hypothesis that supermassive black holes lurked at the centers of most large galaxies after devouring the masses of millions of stars. Today, cosmologists and astrophysicists describe matter as being clumped into enormous clusters and superclusters of galaxies. The quest for these scientists is to explain how the observable universe, consisting of understandable forms of matter and energy, is also immersed in and influenced by mysterious forms of matter and energy, called dark matter and dark energy, respectively.

The study of matter stretches from prehistoric obsidian tools to contemporary research efforts in nanotechnology. States of Matter provides 9th- to 12th-grade audiences with an exciting and unparalleled adventure into the physical realm and applications of matter. This journey in search of the meaning of substance ranges from everyday "touch, feel, and see" items (such as steel, talc, concrete, water, and air) to the tiny, invisible atoms, molecules, and subatomic particles that govern the behavior and physical characteristics of every element, compound, and mixture, not only here on Earth, but everywhere in the universe.

Today, scientists recognize several other states of matter in addition to the solid, liquid, and gas states known to exist since ancient times. These include very hot plasmas and extremely cold Bose-Einstein condensates. Scientists also study very exotic forms of matter, such as liquid helium (which behaves as a superfluid does), superconductors, and quark-gluon plasmas. Astronomers and astrophysicists refer to degenerate matter when they discuss white dwarf stars and neutron stars. Other unusual forms of matter under investigation include antimatter and dark matter. Perhaps most challenging of all for scientists in this century is to grasp the true nature of dark energy and understand how it influences all matter in the universe. Using the national science education standards for 9th- to 12th-grade readers as an overarching guide, the States of Matter set provides a clear, carefully selected, well-integrated, and enjoyable treatment of these interesting concepts and topics.

The overall study of matter contains a significant amount of important scientific information that should attract a wide range of 9th- to 12th-grade readers. The broad subject of matter embraces essentially all fields of modern science and engineering, from aerodynamics and astronomy, to medicine and biology, to transportation and power generation, to the operation of Earth's amazing biosphere, to cosmology and the explosive start and evolution of the universe. Paying close attention to national science education standards and content guidelines, the author has prepared each book as a well-integrated, progressive treatment of one major aspect of this exciting and complex subject. Owing to the comprehensive coverage, full-color illustrations, and numerous informative sidebars, teachers will find the States of Matter to be of enormous value in supporting their science and mathematics curricula.

Specifically, States of Matter is a multivolume set that presents the discovery and use of matter and all its intriguing properties within the context of science as inquiry. For example, the reader will learn how the ideal gas law (sometimes called the ideal gas equation of state) did not happen overnight. Rather, it evolved slowly and was based on the inquisitiveness and careful observations of many scientists whose work spanned a period of about 100 years. Similarly, the ancient Greeks were puzzled by the electrostatic behavior of certain matter. However, it took several millennia until the quantified nature of electric charge was recognized. While Nobel Prize–winning British physicist Sir J. J. (Joseph John) Thomson was inquiring about the fundamental nature of electric charge in 1898, he discovered the first subatomic particle, which he called the electron. His work helped transform the understanding of matter and shaped the modern world. States of Matter contains numerous other examples of science as inquiry, examples strategically sprinkled throughout each volume to show how scientists used puzzling questions to guide their inquiries, design experiments, use available technology and mathematics to collect data, and then formulate hypotheses and models to explain these data.

States of Matter is a set that treats all aspects of physical science, including the structure of atoms, the structure and properties of matter, the nature of chemical reactions, the behavior of matter in motion and when forces are applied, the mass-energy conservation principle, the role of thermodynamic properties such as internal energy and entropy (disorder principle), and how matter and energy interact on various scales and levels in the physical universe.

The set also introduces readers to some of the more important solids in today's global civilization (such as carbon, concrete, coal, gold, copper, salt, aluminum, and iron). Likewise, important liquids (such as water, oil, blood, and milk) are treated. In addition to air (the most commonly encountered gas here on Earth), the reader will discover the unusual properties and interesting applications of other important gases, such as hydrogen, oxygen, carbon dioxide, nitrogen, xenon, krypton, and helium.

Each volume within the States of Matter set includes an index, an appendix with the latest version of the periodic table, a chronology of notable events, a glossary of significant terms and concepts, a helpful list of Internet resources, and an array of historical and current print sources for further research. Based on the current principles and standards in teaching mathematics and science, the States of Matter set is essential for readers who require information on all major topics in the science and application of matter.

Acknowledgments

I wish to thank the public information and/or multimedia specialists at the U.S. Department of Energy (DOE), including those at headquarters and all the national laboratories; the U.S. Department of Defense (DOD) including the individual armed services; the National Institute of Standards and Technology (NIST) within the U.S. Department of Commerce (DOC); the U.S. Department of Agriculture (USDA); the National Aeronautics and Space Administration (NASA), including its centers and astronomical observatory facilities; the National Oceanic and Atmospheric Administration (NOAA) of the DOC; and the U.S. Geological Survey (USGS) within the U.S. Department of the Interior (DOI), for the generous supply of technical information and illustrations used in the preparation of this book set. Also recognized here are the efforts of Frank Darmstadt and other members of the Facts On File team, whose careful attention to detail helped transform an interesting concept into a polished, publishable product. The continued support of two other special people must be mentioned here. The first individual is my longtime personal physician, Dr. Charles S. Stewart III, whose medical skills allowed me to successfully work on this interesting project. The second individual is my wife, Joan, who for the past 45 years has provided the loving and supportive home environment so essential for the successful completion of any undertaking in life.

Introduction

The history of civilization is essentially the story of the human mind understanding the *energy* content of matter. *Energy of Matter* presents many of the most important intellectual achievements and technical developments that led people to more efficiently use matter's energy content. Readers will discover how the ability of human beings to relate the microscopic (atomic level) behavior of matter to readily observable macroscopic properties (such as density, pressure, and temperature) has transformed the world.

Supported by a generous quantity of full-color illustrations and interesting sidebars, *Energy of Matter* discusses the following major topics: the *big bang,* gravity, the relationship between matter and heat, the important role of *thermodynamics,* chemical energy and explosives, the fossil fuels, electrical energy, and *nuclear energy.* Special topics include: the steam engine, wind turbines, geothermal energy systems, the global hydrogen economy, and *dark energy.* Significant breakthroughs in science usually involve inquisitive people and precipitating historical events. This book emphasizes the historical context in which major energy development milestones occurred.

The origin and nature of both matter and energy have perplexed human beings from the dawn of time. The story of energy and its relationship to matter starts with an event called the big bang—a widely accepted theory in contemporary cosmology concerning the origin and evolution of the universe. According to the big bang cosmological model, about 13.7 billion years ago there was an incredibly powerful explosion that started the present universe. Before this ancient explosion, matter, energy, space, and time did not exist. All of these physical phenomena emerged from an unimaginably small, infinitely dense object that scientists call the initial singularity. Immediately after the big bang event, the intensely hot universe, which consisted of pure energy, began to expand and cool. As the universe cooled, matter began to form. Initially, matter consisted of a primeval *quark*-gluon *plasma.* This book describes how the transformation of matter into energy, and vice versa, continues to the present day.

Energy of Matter introduces the reader to the classical (macroscopic) scientific view of energy. Scientists have traditionally defined energy as an ability to do work and have divided energy into two basic categories: *kinetic energy* and potential energy. Kinetic energy is the energy contained or exhibited by matter in motion. Potential energy represents energy stored in a material body or system as a consequence of its position, state, or shape. A coiled spring is an example of potential energy. Scientists often treat chemical energy, nuclear energy, electrical energy, and gravitational energy as various forms of potential energy.

Throughout most of human history, the process of understanding energy has been a very gradual one. The discovery and use of fire in distant prehistoric times represents the first major milestone. Then, several thousand years ago, people learned how to use fire (energy) to process metals and manufacture other new materials, like pottery and bricks. This discovery supported the rise of advanced civilizations all around the Mediterranean Sea and elsewhere. The Bronze Age and the Iron Age represent two important, energy-related, materials science–driven milestones in human history.

About the same time, the development of wind-powered water transportation also influenced the trajectory of human civilization. Early peoples first discovered how to use simple rafts and dugout canoes to travel across inland waterways. Over time, the inhabitants of ancient civilizations learned how to construct ships that employed both human muscles (oars) and wind power (sails) for propulsion. The ancient Egyptians, for example, transported cargo along the Nile River, using various types of barges and sailing ships.

In about 1500 B.C.E., the Phoenicians emerged as the first great maritime trading civilization within the Mediterranean Basin. From the coastal regions of what is now modern Lebanon, Phoenician sailors traveled across the Mediterranean in well-designed, human-powered sailing vessels, referred to by naval architects as biremes. The bireme had two sets of oars on each side of the ship and a large square sail. The ship's name results from a combination of "bi" (meaning two) and "reme" (meaning oars). The Phoenicians were not only master shipbuilders and skilled traders; they also developed the early alphabet upon which present-day alphabets are based. The ancient Greeks and later the Romans improved the design of the Phoenician bireme. The trireme (three rows of oars on each side) emerged as the dominant warship of the Mediterranean Basin and remained so for centuries. Then, when the Dark Ages enveloped most of

Western Europe, Viking longships began departing Scandinavian waters and prowling across the northern Atlantic Ocean. Looking for trade or plunder, Norse sailors ventured far up many great European rivers and made daring sorties into the Mediterranean Sea.

In 1807, a major breakthrough in water transport took place, when the American engineer and inventor Robert Fulton (1765–1815) inaugurated commercial steamboat service. On August 14, 1807, Fulton's steamboat, called the *Clermont,* made its journey up the Hudson River from New York City to Albany, demonstrating the great potential of steam-powered ships. A similar milestone in transportation occurred on January 17, 1955, when the world's first nuclear-powered ship, the submarine USS *Nautilus* (SSN-571), initially put to sea. As the *Nautilus* departed, its captain sent back the historic message: "Underway on nuclear power."

As discussed in *Energy of Matter,* Sir Isaac Newton's laws of motion represent the three fundamental postulates that form the basis of classical mechanics. He formulated these laws in about 1685, while studying the motion of the planets around the Sun. In 1687, Newton (1642–1727) presented his work to the scientific community in *Philosophiae Naturalis Principia Mathematica (Principia).* This influential document represents the capstone of the Scientific Revolution.

In 1775, the Scottish inventor James Watt (1736–1819) entered into a business partnership with the British entrepreneur Matthew Boulton (1728–1809) to produce and install steam engines. Constantly being improved, the Watt steam engine soon dominated mechanical power production all over the United Kingdom and eventually throughout much of Europe and North America. The First Industrial Revolution acted as the major stimulus for improvements in energy science and the production of advanced machines. Occurring in the late 18th and early 19th centuries, it was a time of enormous cultural, technical, and socioeconomic transformations.

British engineers and business entrepreneurs led the charge by developing steam power (fueled by coal). They began using steam engines in manufacturing (especially in the textile industry) and as propulsive devices for more efficient overland transportation systems (that is, railway trains). These technical innovations were soon followed in the early 19th century by the development of all-metal machine tools. As scientists and engineers labored to make steam engines more efficient, they began to explore the nature of heat and its relationship to mechanical work. They also used the scientific method to carefully investigate the thermal properties of matter, giving rise to thermodynamics. Steam not only powered the First Industrial Revolution, it also powered an amazing intellectual revolution in the 19th century—one involving materials science, fluid mechanics, thermodynamics, and the rebirth of atomic theory.

This trend continued well into the Second Industrial Revolution (roughly from 1871 to 1914), when great developments within the oil, steel, chemical, and electrical industries occurred. Mass production of consumer goods—including packaged foods and beverages, clothing, and automobiles (for personal transport)—took place during this period. There is not a clean and crisp demarcation line between the First and Second Industrial Revolutions. But engineering developments in the latter period involved the expanded use of fossil fuel (especially coal and petroleum) as the prime energy source and of electricity as a revolutionary new energy carrier. Scientific discoveries during the Second Industrial Revolution also set the stage for the rise in modern physics, which (in turn) caused the next two technology-based revolutions in human history—the Nuclear Age and the Information Technology (or Digital) Revolution.

One of matter's most interesting and important energetic properties is electromagnetism. In the mid to late 18th century, scientists such as Benjamin Franklin (1706–90), began exploring the fundamental nature of electricity. *Energy of Matter* shows how their pioneering efforts were quickly amplified by a new generation of scientists, like André-Marie Ampère (1775–1836), Charles-Augustin de Coulomb (1736–1806), and Count Alessandro Volta (1745–1827). The invention of the first electric battery (called Volta's electric pile) enabled the start of the Second Industrial Revolution, several decades later. Soon laws governing the flow of electric currents were developed. The British experimental physicist Michael Faraday (1791–1867) and his American counterpart, Joseph Henry (1797–1878), independently discovered the physical principles behind two of the most important electric-powered machines in modern civilization: the electric generator and the electric motor. Later in the 19th century, the Scottish scientist James Clerk Maxwell (1831–79) presented a comprehensive set of equations that theoretically described electromagnetism. Maxwell's brilliant work revolutionized both physics and the practice of engineering. *Energy of Matter* also describes how late 19th-century inventors, such as Thomas Edison (1847–1931) and Nikola Tesla (1870–1943), applied electromagnetism to numerous new devices, which gave humankind surprising new power and comforts.

In 1905, while working for the Swiss federal patent office in Berne, the theoretical physicist Albert Einstein (1879–1955) introduced his famous mass-energy formula, expressed as $E = mc^2$. This famous formula expresses the energy equivalent of matter and vice versa. *Energy of Matter* explains how, among its many important physical insights, Einstein's equation was the key that scientists needed to understand energy release in such important *nuclear reactions* as *fission, fusion*, radioactive decay, and matter-*antimatter* annihilation.

This book reveals how scientists explored the atomic *nucleus* and discovered ways to harvest the incredible amounts of energy hidden within. Understanding the nuclear *atom* forever changed the course of human civilization. *Energy of Matter* explains how the modern understanding of matter on the smallest, or quantum, scale emerged in the 20th century. Especially remarkable progress took place in the decades immediately following World War II. Theoretical physicists, like Murray Gell-Mann (1929–), examined the avalanche of data concerning fundamental particles and then suggested that supposedly elementary particles like *neutrons* and *protons* actually had structure and smaller particles (called quarks) within them. These activities supported the maturation of *quantum mechanics* and resulted in the emergence of the Standard Model of Fundamental Particles and Interactions.

This book explains how the phenomenon of dark energy is helping scientists understand the functioning of the universe. Astrophysical observations indicate that the universe has been expanding ever since the big bang, but at varying rates of acceleration due to a continual cosmic tug of war between the mysterious pushing force of dark energy and the gravitational pulling force of matter—both ordinary matter and *dark matter*. Scientists have attempted to explain the observed increased rate of expansion of the universe by postulating the notion of a special force, called dark energy, which opposes gravity and allows the universe to expand at an ever more rapid rate.

Energy of Matter has been carefully designed to help any student or teacher who has an interest in the overall energy mysteries of matter, who wants to understand what the relationship between matter and energy is, who seeks to understand how scientists measure and characterize energy, or who wishes to understand how the knowledge and use of energy shaped the course of human civilization. The back portion of the book contains an appendix with a contemporary periodic table, chronology, glossary, and an array of historical and current sources for further research. These should prove especially helpful for readers who need additional information on specific terms, topics, and events.

The author has carefully prepared the volume so that a person familiar with SI units (the international language of science) will have no difficulty understanding and enjoying its contents. The author also recognizes that there is a continuing need for some students and teachers in the United States to have units expressed in the United States (American) Customary system of units. Wherever appropriate, both unit systems appear side by side. An editorial decision places the American units first, followed by the equivalent SI units in parentheses. This format does not imply the author's preference for American units over SI units. Rather, the author strongly encourages all readers to take advantage of the particular formatting arrangement to learn more about the important role that SI units play within the international scientific community.

CHAPTER 1

Understanding Energy

The ability of human beings to relate the microscopic (atomic level) energetic behavior of matter to readily observable macroscopic properties (such as temperature, pressure, and density) transformed science and engineering. While ancient Greek philosophers such as Aristotle (384–322 B.C.E.) speculated about why matter moved, the great engineers of antiquity such as Archimedes of Syracuse (ca. 287–212 B.C.E.) used energy-related phenomena effectively—without any real scientific understanding of what energy was. The initial breakthroughs in accurately describing and measuring energy occurred in the 17th century, as part of the Scientific Revolution. This chapter uses modern technology examples and historical anecdotes to describe some of humankind's early efforts to understand energy. The chapter concludes with a brief summary of the present interpretation of energy. Subsequent chapters enlarge and enrich the brief summary presented here.

EARLY CONCEPTS OF MATTER AND ENERGY

Matter is a much easier concept to understand than energy. When a caveman picked up a rock, he recognized it had size, shape, color, and substance (mass). When he threw that rock up into the air, he observed that it climbed to a certain height, paused briefly, and then came back to the ground. In all probability, no prehistoric philosopher interpreted the

The ascent of a powerful rocket into space is a vivid demonstration of potential energy within the chemical fuel being transformed into kinetic energy of the flight vehicle. This is NASA's space shuttle *Discovery* on its way to the *International Space Station* on July 26, 2005. The STS-114 mission involved a seven-member astronaut crew. *(NASA)*

free-fall trajectory of a tossed stone as an exchange between kinetic energy and potential energy (neglecting, for the moment, aerodynamic friction). That important discovery would be made in the 17th century by the brilliant Italian physicist and mathematician, Galileo Galilei (1564–1642).

It was because a very large rock crashed into Earth some 65 million years ago that small mammals were able to emerge from the shadows of a world previously ruled for many millions of years by giant reptiles called dinosaurs. Many scientists accept the physical evidence, which suggests that a very large asteroid (about six miles [10 km] in diameter) plunged through the atmosphere of prehistoric Earth and hit with enormous force in the shallow, tropical seas near the Yucatán Peninsula of modern Mexico. This intensely energetic cosmic collision generated blast waves and spewed hundreds of billions of tons of sulfur-rich soil into Earth's atmosphere, producing a worldwide blackout and freezing temperatures.

An artist's rendering that shows pterodactyls, flying reptiles with wingspans of up to 50 feet (15 m), as they glide above the low tropical clouds near the Yucatán Peninsula about 65 million years ago. In the background is a giant asteroid (about six miles [10 km] in diameter), approaching the Chicxulub impact site. The cataclysmic ancient collision produced a drastic change in global conditions—causing the extinction of the dinosaurs and allowing the rise of mammals, including (eventually) human beings. *(NASA; artist, Donald E. Davis)*

Scientists speculate that these harsh environmental conditions persisted for at least a decade. They further suggest that global disruption of the biosphere promoted the extinction of the dinosaurs and many other species. Out of the environmental chaos of this ancient extinction event small mammals slowly emerged, began to evolve, and eventually filled available ecological niches.

The Lower Paleolithic period extends from approximately 2 million years ago until about 100,000 B.C.E. and represents the earliest and longest specific period of human development. During this period, early hunter-gatherers learned to use simple stone tools for cutting and chopping. Crude handheld stone axes represented a major advancement. In the Middle Paleolithic period (from about 100,000 B.C.E. to about 40,000 B.C.E.), Neanderthal man lived in caves, learned to control fire, and employed improved stone tools for hunting. These nomadic people also learned how to use bone needles to sew furs and animal skins into body coverings. Archaeological evidence suggests that early peoples began to employ fire as a powerful tool, not only for warmth and illumination, but also to repel dangerous animals, to herd and hunt desirable game, and even to set controlled burns that encouraged the growth of game-attracting, grassy vegetation. Although human beings used fire for untold millennia, it was not until the 19th century that the energy-releasing process of combustion achieved adequate scientific interpretation. Prior to that time, people often assigned fire elemental status, as did the ancient Greeks.

During the Upper Paleolithic period (from about 40,000 B.C.E. to approximately 10,000 B.C.E.), Cro-Magnon man arrived on the scene and displaced Neanderthal man. Cro-Magnon tribal clans initiated more organized and efficient hunting and fishing activities with improved tools and weapons, including carefully sharpened obsidian and flint blades. A great variety of finely worked stone tools, better sewn clothing, the first human-constructed shelters, and jewelry (such as bone or ivory necklaces) also appeared. The shaman of Cro-Magnon clans painted colorful animal pictures on the walls of caves. The ancient cave paintings that survive provide scientists some insight into the rituals of Cro-Magnon hunting cultures.

The first important step leading to human understanding of the energy of matter was the discovery and use of fire in prehistoric times. A companion intellectual breakthrough by early humans involved the development and use of very simple tools, such as a stick, a sharp rock, and similar objects. These developments were extremely gradual; when they

The discovery and use of fire ultimately led to the rise of human civilization. This artist's rendering shows Earth (Europe) about 60,000 years ago. Without understanding the process of combustion, members of this Neanderthal hunter-gather band prepare a meal, while huddling around a fire for warmth and security. Woolly mammoths and other Pleistocene animals appear in the background. The bright red dot in the night sky is the planet Mars. *(NASA; artist, Randii Oliver)*

did occur, they typically led to a division of labor within ancient hunting and gathering societies. The survival of a prehistoric clan in the late Upper Paleolithic period not only depended upon the skill of its hunters and gatherers, but also upon the expertise of its fire-keepers and toolmakers.

The Mesolithic period (also called the Middle Stone Age) featured the appearance of better cutting tools, carefully crafted stone points for spears and arrows, and the bow. Depending on the specific geological region, the Mesolithic period began about 12,000 years ago. This era of human development continued until replaced by another important technical and social transformation anthropologists call the Neolithic Revolution.

Scientists believe that humans domesticated the dog during the Mesolithic period and then began using the animal as a hunting companion. This prehistoric human-canine bond continues to this very day.

The Neolithic Revolution (New Stone Age) was the incredibly important transition from hunting and gathering to agriculture. As the last ice age ended about 12,000 years ago, various prehistoric societies in the Middle East (the Nile Valley and Fertile Crescent) and in parts of Asia independently began to adopt crop cultivation. Since farmers tend to stay

EARTHMOVING EXPERIMENT

The United States Atomic Energy Commission (AEC), now the Department of Energy (DOE), established the Plowshare Program in 1958 as a research and development activity to explore the technical and economic feasibility of using nuclear explosives for industrial (civilian) applications. The reasoning was that the relatively inexpensive energy available from nuclear explosions could prove useful for a wide variety of peaceful purposes. The program's prime objective gave rise to the term *peaceful nuclear explosive* (PNE). Between December 1961 and May 1973, the United States conducted 27 Plowshare nuclear explosive tests, consisting of 35 individual detonations. The Plowshare Program ended in 1975. The former Soviet Union engaged in a similar PNE program.

In concept, industrial applications resulting from the use of nuclear explosives can be divided into two broad categories: large-scale excavations and quarrying, where the energy of the nuclear explosion is used to break up and/or move rock and soil, and underground engineering, where the energy released from deeply buried nuclear explosives increases the permeability and porosity of the rock by massive breaking and fracturing. The design of the nuclear explosive and its detonation location (ground zero) underground minimized the amount of radiation released to the environment by the blast.

Suggested excavation applications within the Plowshare Program included canals, harbors, highway and railroad cuts through mountains, open-pit mining, and the construction of dams. Underground PNE applications included the stimulation of natural gas production, the preparation of suitable ore bodies for in situ leaching, the creation of underground zones of fractured oil shale for in situ retorting, and the formation of underground natural gas and petroleum storage reservoirs.

in one place, the early peoples involved in the Neolithic Revolution began to establish semipermanent and then permanent settlements. Historians often identify this period as the beginning of human civilization. The Latin word *civis* means citizen or a person who inhabitants a city.

Throughout the history of civilization, one of the most energy-demanding activities associated with the construction of individual habitats, cities, roads, and defensive structures (such as walls) involved moving enormous quantities of soil and rocks. Hauling "dirt" to, from, or around

As part of the Plowshare Program, the United States conducted a spectacular excavation experiment in 1962 called Sedan. The experiment took place in alluvium soil at the Nevada Test Site. The objective of this PNE test was to determine the feasibility of using nuclear explosions for large-scale excavation projects, such as canal construction. Scientists detonated a 104-kiloton yield nuclear device buried at a depth of 637 feet (194 m) below the surface in alluvium soil. The explosion created a huge crater 1,280 feet (390 m) in diameter and 330 feet (100 m) deep. The excavating explosion moved an estimated 201 million cubic feet (5.7×10^6 m^3) of alluvium soil.

The Sedan peacetime nuclear explosion excavating test at the Nevada Test Site on July 6, 1962. In a single blast, this 104-kiloton yield subsurface detonation moved an estimated 201 million cubic feet (5.7×10^6 m^3) of alluvium soil and created a huge crater 1,280 feet (390 m) in diameter and 330 feet (100 m) deep. *(DOE)*

One kiloton (kT) is a unit of energy defined as being equivalent to 3.98×10^9 Btu (4.2×10^{12} joules). Scientists use this unit to describe the energy released in a nuclear detonation. It is approximately equal to the amount of energy that would be released by the explosion of 1,000 metric tons (that is, one kiloton) of the chemical high explosive trinitrotoluene (TNT).

a construction site still remains an integral part of civil engineering and the business of building homes, roads, factories, scientific laboratories, skyscrapers, and airports.

The early Greek philosopher Thales of Miletus (ca. 624–545 B.C.E.) was the first European thinker to suggest a theory of matter. About 600 B.C.E., he postulated that all substances came from water and would eventually turn back into water. Thales may have been influenced in his thinking by the fact that water assumes all three commonly observed states of matter: solid (ice), liquid (water), and gas (steam or water vapor). The evaporation of water under the influence of fire or sunlight could have provided Thales the notion of recycling matter.

Anaximenes (ca. 585–525 B.C.E.) was a later member of the school of philosophy at Miletus. He followed in the tradition of Thales by proposing one primal substance as the source of all other matter. For Anaximenes that fundamental substance was air. In surviving portions of his writings, he suggested, "Just as our soul being air holds us together, so do breathe and air encompass the whole world." He proposed that cold and heat, moisture and motion made air visible. He further stated that when air was rarified it became diluted and turned into fire. In contrast, the winds for Anaximenes represented condensed air. If the condensation process

A distinctive condensation cone develops around a U.S. Navy F/A-18 Hornet as it breaks the sound barrier. *(U.S. Navy)*

A close-up look at a raging wildfire—this scene depicts trees torching within Yellowstone National Park on September 4, 1988. *(NPS)*

continued the result was water, with further condensation resulting in the primal substance (air) becoming earth and then stones.

The Greek pre-Socratic philosopher Empedocles (ca. 495–435 B.C.E.) lived in Acragas, a Greek colony in Sicily. In about 450 B.C.E., he wrote the poem *On Nature.* In this lengthy work (of which only fragments survive), he introduced his theory of the universe in which all matter is made up of four classical elements: earth, air, water, and fire—that periodically combine and separate under the influence of two opposing forces (love and strife). According to Empedocles, fire and earth, when combined, produce dry conditions; earth blends with water to form cold; water combines with air to produce wet; and air and fire when combined form hot.

PARADOX OF FIRE

From a scientific perspective, fire is a rapid, persistent chemical reaction that releases heat and light. Fire involves the exothermic (energy-releasing) combination of a combustible substance (fuel) and an oxidant (typically, the gaseous oxygen found in the atmosphere). Chemists define *combustion* as a chemical change, especially an oxidation reaction, accompanied by the release of heat and light.

Fire is both lethal and the key to human survival and progress. Paleoclimatologists speculate that Earth's primitive atmosphere most likely resembled the gases released during volcanic eruptions, namely water vapor (H_2O), carbon dioxide (CO_2), sulfur dioxide (SO_2), carbon monoxide (CO), sulfur (S_2), chlorine (Cl_2), nitrogen (N_2), hydrogen (H_2), methane (CH_4), and ammonia (NH_3). There was no free oxygen (O_2). So, for about 90 percent of this planet's approximately 4.5 billion-year history, there was no fire on its surface. It was only about 400 million years ago that Earth's atmosphere attained its current level of free oxygen (about 21 percent by volume). Plants on the land (biomass) provided fuel for fire; photosynthesis provided a sufficient concentration of free oxygen in the air to support combustion; and lightning provided the natural spark. Prehistoric wildfires raged unchecked, until quenched by rainstorms, limited by natural barriers, or self-extinguished due to fuel depletion. Nature remained totally in charge of fire for the next 398 million or so years.

About 1.5 million years ago, prehistoric humans learned about fire, how to make it, and eventually how to control it. Fire gave early humans unprecedented

The notion of energy (here, fire), as an elemental substance, persisted in Western civilization in various theories up through the mid-19th century.

In about 430 B.C.E., the early Greek philosopher Democritus (ca. 460–370 B.C.E.) elaborated upon the atomic theory of matter—expanding upon a theory initially suggested by his teacher Leucippus (fifth century B.C.E.). Democritus emphasized that all things consist of changeless (eternal), indivisible, tiny pieces of matter, called atoms. According to Democritus, different materials consisted of different atoms, which interacted in specific ways to produce the particular properties of a specific material. Some types of solid matter consisted of atoms with hooks, so they could attach to each other. Other materials, like water and air, consisted

power over their world. A campfire's warmth and light provided protection; cooking expanded the range and quality of food available for survival. Early hunter-gatherers even learned how to use fire to clear the land of scrub vegetation, in order to promote the growth of game-attracting vegetation. As the first civilizations emerged, higher temperature fires (based on the use of kilns and charcoal) enabled the production of pottery and metals. By accident or through acts of violence in warfare, fire also killed people and destroyed early settlements. Throughout most of human history, wood served as the primary fuel for fires. The arrival of more efficient steam engines as part of the First Industrial Revolution encouraged expanded use of coal in the late 18th century.

It was just in the last 200 years that people learned how to efficiently harvest most of the chemical energy stored in fossil fuels (coal, petroleum, and natural gas). They discovered how to produce large quantities of mechanical energy (using the steam engine), how to generate large amounts of electricity (using fossil fuel power plants), and how to travel swiftly through the air (using hydrocarbon-fueled jet engines). Scientists now understand that the complete combustion of carbon releases both thermal energy and carbon dioxide. They also recognize that the incomplete combustion of carbon results in the release of carbon monoxide (CO), soot (primarily unburned particles of carbon), and other atmospheric pollutants.

The use of fire enabled the rise of a global civilization. Many scientists express concern that the current, large-scale use of carbon-fueled fire could so degrade the environment as to hurl modern civilization into an irreversible, downward spiral.

This 1961 stamp from Greece honors the early Greek philosopher Democritus (ca. 460–370 B.C.E.) *(courtesy of the author)*

of large, round atoms that moved smoothly past each other. What is remarkable about the ancient Greek theory of atomism is that it tried to explain the great diversity of matter found in nature with just a few basic ideas tucked into a relatively simple theoretical framework. The atomistic theory of matter fell from favor when the more influential philosopher Aristotle rejected the concept.

Starting in about 340 B.C.E., Aristotle embraced and embellished the theory of matter originally proposed by Empedocles. Within Aristotelian cosmology, planet Earth is the center of the universe. Everything within Earth's sphere is composed of a combination of the four basic elements: earth, air, water, and fire. Aristotle suggested that objects made of these four basic elements are subject to change and move in straight lines. Aristotle coined the term *energia* (ενεργια) to describe the motion or activity of matter. He developed this new word from the Greek words: *energos* (ενεργος), meaning active, and *ergon* (εργον), meaning work. For Aristotle, energia was intimately connected to the metaphysical notion of a potentiality transforming into an actuality. While energy traces its linguistic origin to Aristotle's *energia,* the modern scientific interpretation of this word did not appear until the early 19th century. In 1807, the British physician and scientist Thomas Young (1773–1829) became the first person to use *energy* in its current scientific context.

Aristotle further speculated that heavenly bodies were not subject to change and moved in perfect circles. He stated that beyond Earth's sphere there was a fifth basic element, which he called *aether* (αιθηρ)—a pure form of air that was not subject to change. Finally, the Greek philosopher declared that he could analyze all material things in terms of matter (substance) and form (essence). Aristotle's ideas about the nature of matter and the structure of the universe dominated thinking in Europe for centuries, until finally displaced during the Scientific Revolution. For example, the concept of the immutable (unchanging) heavens permeated the practice of astronomy in Western civilization until Galileo Galilei assaulted and dismantled the erroneous premise in the early 17th century.

Galileo's discovery of sunspots clearly demonstrated that the Sun was subject to change.

HERON OF ALEXANDRIA

Heron of Alexandria (first century, ca. 20–80 C.E.) was the last of the great Greek engineers of antiquity. Also called Hero, he invented many clever mechanical devices, including the aeolipile, the device for which he is most commonly remembered. The aeolipile is a spinning, steam-powered spherical apparatus that demonstrated the action-reaction principle, which forms the basis of Sir Isaac Newton's third law of motion and the operating principle for modern reaction devices, such as gas turbines and jet engines. Scientists regard the aeolipile as the earliest known example of the steam turbine—a mechanical device capable of transforming the energy content of steam into rotary motion.

Not much has survived from antiquity about the personal life of Heron. Historians estimate that the Greek inventor was born in about 20 C.E., because his writings indicate that he observed a lunar eclipse, which was observable in Alexandria in 62 C.E. Heron had a strong interest in simple machines, mechanical mechanisms (like gears), and fluid (hydraulic and pneumatic) systems. His inventions and publications reflect the influence of an earlier Greek engineer, called Ctesibius of Alexandria (ca. 285–222 B.C.E.). Several of Heron's works have survived, including *Pneumatics* (written about 60 C.E.), *Automata, Mechanics, Dioptra,* and *Metrics.*

His most familiar invention is the aeolipile. He placed a hollow metal sphere on pivots over a charcoal grill–like device. When water placed inside the metal sphere was heated over the grill, steam formed and escaped through the tubes, which acted like crude nozzles. The sphere would spin freely on pivots, as steam escaped from two small opposing tubes connected to the sphere. This whirling sphere delighted children and became a popular toy. For some inexplicable reason, Heron never connected the action-reaction principle exhibited by the aeolipile with the basic concept of steam-powered devices for performing useful mechanical work.

The aeolipile is an example of a clever device invented well ahead of its time. Such devices sometimes need to be "rediscovered" or "reinvented" decades or even centuries later, when the social, economic, and/or technical conditions are just right for full engineering development and application. Since the aeolipile embodies the action-reaction principle, it is the technical ancestor of modern power turbines, jet engines, and rocket

TURBINES

A turbine is a machine that converts the energy of a fluid stream into mechanical energy of rotation. The working fluid used to drive a turbine can be gaseous or liquid. There are many types of turbines. A highly compressed gas drives an expansion turbine; hot combustion gases drive a gas (or combustion) turbine; steam (or other vapor) drives a steam (or vapor) turbine; water drives a *hydraulic* turbine; and wind spins a wind turbine (or windmill).

Except for wind and water-powered turbines, a modern turbine typically consists of two sets of curved blades (or vanes) alongside each other. One set of vanes is fixed and the other set can move. Engineers space the moving vanes (called the rotor) around the circumference of a cylinder, which can rotate about a central shaft. They then attach the fixed set of vanes (called the stator) to the inside casing than encloses the rotor, or moving portion of the turbine.

In order to efficiently extract energy from the working fluid, gas and steam turbines usually contain a series of successive stages. Each stage consists of a set of fixed (stator) and moving (rotor) blades. The pressure of the working fluid decreases, as it passes from stage to stage. Engineers increase the overall diameter of each successive stage to maintain a constant torque (or rotational effect) as the working fluid expands and loses pressure and energy.

Power system engineers design a combustion turbine to start quickly in order to meet the demand for electricity during peak periods. These turbines normally operate with natural gas as the fuel, although low-sulfur content fuel oil may also be used. The combustion turbine functions somewhat like a jet engine—however, no propulsive thrust is produced. Specifically, the system draws in ambient air at the front, compresses it, mixes the higher-pressure compressed air with fuel, and then combusts (ignites) the fuel-air mixture. As the hot combustion gases expand through the turbine, they spin (rotate) the turbine, which is located on a common shaft that is also connected to the compressor and to an electricity-producing generator. (In a jet engine, the combustion gases pass through a turbine and then expand through an exit nozzle to produce thrust.) After the hot gases pass through the power turbine and transform most of their energy content into the mechanical energy of rotary motion, the spent gases exhaust safely to the surroundings. The transformer at the power plant accommodates the long-distance transmission of the electricity through the power grid.

engines. In the 20th century, steam turbines helped electrify the world; jet engines allowed modern military and commercial aircraft to swiftly reach distant points around the globe; and powerful rockets enabled the exploration of the solar system.

Despite his oversight regarding useful applications of the aeolipile, Heron remained a skilled engineer and creative inventor. He developed a variety of feedback-control devices that used fire, water, and compressed air in different combinations. He designed a machine for threading wooden screws and constructed an automated puppet theater. He also receives credit for creating an early odometer, a primitive form of analog computer (involving gears, spindles, weights, pegs, trays of sand, and ropes) and a compressed air (pneumatic) fountain.

HARNESSING ENERGY FOR TRANSPORTATION

For most of human history, muscle power (human or animal) served as the major energy source for travel across Earth's surface or for hauling loads. People in early civilizations learned to domesticate certain animals, like horses, oxen, and mules. They rode these domesticated animals for transportation or else used them to pull heavy loads, including wagons and chariots.

A draft horse (or workhorse) is any breed of horse capable of drawing heavy loads. Draft horses originated in central Europe, where their domestication preceded the Roman invasion. All breeds are extremely large and noted for their strength and endurance.

To market his improved steam engines in the late 18th century, the Scottish engineer James Watt (1736–1819) developed the term *horsepower* (hp). Using horsepower as a figure of merit, prospective customers could compare how much power a particular steam engine developed versus the rate of work being performed by the draft horses and ponies commonly used in British mining operations. (Engineers define power as the rate of energy usage or energy transfer.) Watt observed how much mass a typical draft animal could lift in a minute. (In American customary units, this represented about 22,000 foot-pounds-force per minute [ft-lbf/min].) Increasing the basic amount of power (that is, work per unit time) by an arbitrary 50 percent, Watt arrived at 33,000 ft-lbf/min for one horsepower. Although not regarded as an internationally accepted scientific unit, horsepower remains well entrenched in modern society. American automotive engineers define one horsepower as 550 foot-pounds-force per

The Budweiser Clydesdales, among the most recognizable advertising symbols, during a public appearance saluting the military in 2008 at Offutt Air Force Base, Nebraska; originating from the Clyde valley of Scotland, the Clydesdales are a powerful breed of draft horse. *(U.S. Air Force)*

second (ft-lbf/s). In *SI units* (*International System of Units*, from the French Le Système International d'Unités), this level of power corresponds to approximately 746 joules per second (J/s) or watts (W). Subsequent chapters will discuss how scientists developed a consistent set of units to measure energy and power.

The development of wind-powered water transportation significantly influenced the trajectory of human civilization. Early peoples first discovered how to use simple rafts and dugout canoes to travel across inland waterways. Over time, the inhabitants of ancient civilizations learned how to construct ships that used both human muscle (oars) and wind power (sails) for propulsion. The Egyptians transported cargo along the Nile River, using various types of barges and sailing ships. Then, as now, water transport offered an efficient way of moving large quantities of materials.

In about 1500 B.C.E., the Phoenicians emerged as the first great maritime trading civilization within the Mediterranean Basin. From the coastal regions of what is now modern Lebanon, Phoenician sailors traveled across the Mediterranean in the well-designed, human-powered sailing vessels called biremes. The bireme had two sets of oars on each side of the ship and a large square sail. The ship's name results from a combination of "bi" (meaning two) and "reme" (meaning oars). The Phoenicians were master shipbuilders, skilled traders, and developed the early alphabet upon which present-day alphabets are based. The ancient

The ancient Egyptians pioneered the development of river craft, since the slow-flowing Nile River was ideal for transportation. This 1963 stamp from San Marino depicts an early Egyptian cargo ship, designed to carry agricultural produce. *(courtesy of the author)*

Greeks and later the Romans improved the design of the Phoenician bireme. The trireme (three rows of oars on each side) emerged as the dominant warship of the Mediterranean Basin for centuries. Then, when the Dark Ages enveloped most of western Europe, Viking longships began departing Scandinavian waters and prowling across the northern Atlantic Ocean. Looking for trade or plunder, Norse sailors ventured far up many great European rivers and made daring sorties into the Mediterranean Sea.

As the nations of western Europe began their ambitious programs of worldwide exploration in the 15th century, their military and commercial interests encouraged the design of better sailing ships. For about the next 400 years, a parade of more efficient sailing ships carried commercial products, adventurers, and new ideas to all portions of the globe.

In 1807, a major breakthrough in water transport took place when the American engineer and inventor Robert Fulton (1765–1815) inaugurated commercial steamboat service. On August 14, 1807, Fulton's steamboat, called the *Clermont*, made its journey up the Hudson River from New York City to Albany, demonstrating the great potential of steam-powered ships. A similar milestone in transportation occurred on January 17, 1955, when the world's first nuclear-powered ship, the submarine USS *Nautilus*

(SSN-571), initially put to sea. As the *Nautilus* departed, its captain sent back the historic message: "Underway on nuclear power."

SCIENTIFIC FORMS OF ENERGY

This section provides a brief summary of how scientists and engineers now view the concept of energy. First, they define energy as the capacity to do work. Matter moves and changes because of energy. Everything has energy. One of the great principles of modern science states that matter and energy can neither be created nor destroyed but simply experience transformations. For example, during nuclear fission (splitting) reactions within the core of an operating reactor, very small amounts of matter disappear as heavy nuclei split apart. The amount of energy that appears precisely corresponds to the tiny amount of nuclear mass that has disappeared.

A member of the U.S. Air Force Academy's 2009 intercollegiate baseball team swings at an incoming pitch during a game against the University of Nevada (Las Vegas) at Falcon Field in Colorado Springs. *(U.S. Air Force)*

Traditionally, scientists have found it convenient to divide energy into two broad categories: kinetic energy and potential energy. Kinetic energy is energy associated with motion, while potential energy represents stored energy. A baseball hurled by a pitcher at 95 miles per hour (MPH; 153 km/h) has a significant amount of kinetic energy.

A coiled spring has potential energy. So does the food a person eats. Food satisfies three basic needs of the human body: Food provides the materials needed for the growth and repair of tissue; food provides the collection of chemical compounds that the body uses to regulate various biological activities; and food provides energy. A person's body converts about 80 percent of food's energy content into heat. The remaining amount of food energy supports muscular actions and various biochemical activities within the body. Scientists recognize that the body transforms the potential (chemical) energy contained in food through metabolism—a biological process somewhat analogous to combustion. When burned, one gram of fat yields (on average) about 9.0 kilocalories (38.0 kilojoules [kJ]) of heat; while one gram of carbohydrates yields about 4.0 kcal (16.7 kJ). Fats contain more energy value per unit mass than do carbohydrates.

The calorie (cal) is a unit of energy associated with a now discarded theory of heat called caloric theory (see chapter 4). In the 18th century, scientists such as the French chemist Antoine Laurent Lavoisier (1743–94) defined the calorie as the amount of heat required to raise one gram of water one degree Celsius. (One calorie corresponds to 4.1868 joules or 0.0040 Btu.) In studies of metabolism, American physiologists often use a unit called the kilocalorie—or big Calorie (symbol Cal)—to describe the energy content of food. One big (food) Calorie is equivalent to 1,000 small (heat-related) calories. The World Health Organization (WHO) recommends that food energy be expressed in kilojoules (kJ). In this case, one Cal is equivalent to 4.1868 kJ, or 1.0 kcal. These units can be a bit confusing, especially when a person is trying to compare foods packaged in different countries. In the United States, the energy value of food is usually expressed as big (or large) Calories. A small fresh apple contains about 53 Cal (220 kJ); while an average-sized slice of apple pie contains approximately 277 Cal (1,159 kJ).

Finally, scientists divide energy sources into two general groups: nonrenewable and renewable. Nonrenewable energy sources include the fossil fuels (coal, petroleum, and natural gas), as well the element uranium. Renewable energy resources include solar energy, geothermal energy,

As American as apple pie; on average, each American eats more than 19 pounds (8.6 kg) of apples annually. *(USDA; photographer, Scott Bauer)*

wind energy, hydropower, and biomass energy (such as wood). Scientists and engineers regard electricity as an energy carrier. Like electricity, hydrogen is also considered a secondary source of energy and an energy carrier. Subsequent chapters describe and discuss all of these energy-related topics in more detail.

CHAPTER 2

The Big Bang—Source of All Energy and Matter

About 13.7 billion years ago, an incredibly powerful explosion took place. Called the big bang, this enormous explosion created space and time, as well as all the energy and matter that now exists in the universe. This chapter summarizes the latest scientific understanding about this incredible event. Physicists are still not sure what "banged" or why it "banged." But immediately after the big bang, the universe began to expand, and energy and matter began an incredible series of transformations that have led to the universe in its present state, which includes stars, planets, and intelligent life. This chapter discusses the origin of ordinary (or baryonic) matter within the context of big bang cosmology. A brief mention is also made of the phenomenon of dark energy and how it, along with dark matter, has helped define the currently observed universe.

THE BIG BANG

The big bang is a widely accepted theory in contemporary cosmology concerning the origin and evolution of the universe. According to the big bang cosmologists, about 13.7 billion years ago there was an incredibly powerful explosion that started the present universe. Before this ancient explosion, matter, energy, space, and time did not exist. All of these physical phenomena emerged from an unimaginably small, infinitely dense object, which physicists call the initial singularity. Immediately after the big bang, the intensely hot universe began to expand and cool. As the temperature dropped, energy and matter started to experience several fascinating transformations.

Astrophysical observations indicate that the universe has been expanding ever since the big bang, but at varying rates of acceleration. The continual cosmic tug-of-war between the mysterious pushing force of dark energy and the gravitational pulling force of matter (both ordinary

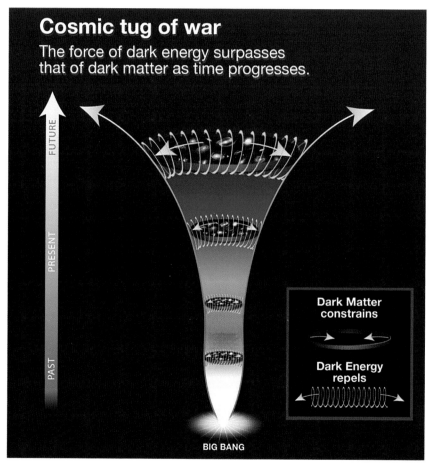

This diagram suggests that the overall history of the universe can be envisioned as a cosmic tug-of-war between the gravitational pull of dark matter (the universe's major form of matter) and the push of dark energy (a mysterious phenomenon of nature). This process began at least 9 billion years ago, well before dark energy gained the upper hand and began accelerating the expansion of the universe. *(NASA, ESA, and A. Feild [STScI])*

matter and dark matter) appears responsible for this variation in rate of acceleration.

The key observation that supported big bang cosmology took place in 1964, when the German-American physicist Arno Allan Penzias (1933–   ) and the American physicist Robert Woodrow Wilson (1936–   ) detected the cosmic microwave background (CMB). Physicists regard the CMB as the remnant radiation signature of an intensely hot young universe. Although the discovery occurred quite by accident, their pioneering work provided the first direct scientific evidence that the universe had an explosive beginning.

Prior to the discovery of the CMB, the majority of scientists were philosophically comfortable with a steady state model of the universe—a model postulating that the universe had no beginning and no end. Within this steady state model, scientists assumed matter was being created continuously (by some undefined mechanism) to accommodate the universe's observed expansion. In 1929, the American astronomer Edwin P. Hubble (1889–1953) proposed the concept of an expanding universe. He suggested that his observations of Doppler-shifted wavelengths of the light from distant galaxies indicated that these galaxies were receding from Earth with speeds proportional to their distance. Following Hubble's discovery, the steady state model became especially popular because it allowed an expanding universe to maintain an essentially unchanging appearance over time.

In 1949, the British astronomer Sir Fred Hoyle (1915–2001), who helped develop and strongly advocated the steady state model, coined the term *big bang.* Hoyle used the term derisively against fellow astronomers who favored big bang cosmology. Unfortunately, Hoyle's derogatory use of the term backfired. The expression immediately gained favor among those competing scientists, who warmly embraced big bang as a clever way for them to succinctly explain the new theory. Detailed observations of the CMB proved that the universe did indeed have an explosive beginning about 13.7 billion years ago. As evidence mounted in favor of big bang cosmology, the vast majority of scientists abandoned the steady state universe model.

Near the end of the 20th century, big bang cosmologists had barely grown comfortable with a gravity-dominated expanding universe model when a major cosmological surprise appeared. In 1998, two competing teams of astrophysicists independently observed that the universe was not just expanding, but expanding at an increasing rate.

Temperature fluctuations of the cosmic microwave background: The average temperature is 2.725 kelvins (K) and the colors represent the tiny temperature fluctuations. Red regions are warmer and blue regions are colder by about 0.0002 K. This full-sky map of the heavens is based on five years of data collection by NASA's *Wilkinson Microwave Anisotropy Probe (WMAP). (NASA/WMAP Science Team)*

Dark energy is the generic name that astrophysicists give to the unknown cosmic force field believed responsible for the acceleration in the rate of expansion of the universe. In the late 1990s, scientists performed systematic surveys of very distant Type Ia (carbon detonation) supernovas. They observed that instead of slowing down (as might be anticipated, if gravitation was the only significant force at work in cosmological dynamics), the rate of recession (that is, the redshift) of these very distant objects appeared to actually be increasing. It was almost as if some unknown force was neutralizing or canceling the attraction of gravity. Such startling observations proved controversial and very inconsistent with the then standard gravity-only models of an expanding universe within big bang cosmology. Today, carefully analyzed and reviewed Type Ia supernova data indicate that the universe is definitely expanding at an accelerating rate.

Physicists do not yet have an acceptable answer as to what phenomenon is causing this accelerated rate of expansion. Some scientists revisited the cosmological constant (symbol Λ) that Einstein inserted into his original general relativity theory to make that theory of gravity describe a static universe. A static universe would be a non-expanding one that had neither a beginning nor an end. After boldly introducing the cosmological constant as representative of some mysterious force associated with empty space capable of balancing or even resisting gravity, Einstein abandoned the idea. Hubble's announcement of an expanding universe provided the intellectual nudge that encouraged his decision. Later in life, Einstein would refer his postulation of a cosmological constant as "my greatest failure."

But Einstein may have been on the right track all along. Physicists are now revisiting Einstein's concept and suggesting that there is possibly a vacuum pressure force (a recent name for the cosmological constant) that is inherently related to empty space, but exerts its influence only on a very large scale. The influence of this mysterious force would have been negligible during the very early stages of the universe following the big bang event but would later manifest itself and serve as a major factor in cosmological dynamics. Since such a mysterious force is neither required nor explained by any of the currently known laws of physics, scientists do not yet have a clear physical interpretation of what this gravity-resisting force means.

DARK MATTER AND DARK ENERGY

It has taken the collective thinking of some of the very best minds in human history many millennia to recognize the key scientific fact that all matter in the observable universe consists of combinations of different atoms drawn from a modest collection of about 100 different chemical elements. (See appendix.) The term *observable universe* is intentionally used here, because scientists are now trying to understand and characterize an intriguing invisible form of matter that they call dark matter.

This book deals primarily with the energetic characteristics and properties of ordinary (or baryonic) matter. Scientists define ordinary matter as matter composed primarily of baryons. Baryons are typically the heavy nuclear particles formed by the union of three quarks, such as protons and neutrons. Almost everything a person encounters in daily life consists of atoms composed of ordinary (or baryonic) matter. (Note that whenever the word *matter* appears in this book, it means ordinary [or baryonic] matter unless specifically indicated otherwise.)

Scientists define dark matter as matter in the universe that cannot be observed directly because it emits very little or no electromagnetic radiation and experiences little or no directly measurable interaction with ordinary matter. They also refer to dark matter as nonbaryonic matter. This means that dark matter does not consist of baryons, as does ordinary matter. So what is dark matter? Today, no one really knows. While not readily observable here on Earth, dark matter nevertheless exerts a very significant, large-scale gravitational influence within the universe.

Starting in the 1930s, astronomers began suspecting the presence of dark matter (originally called missing mass) because of the observed velocities and motions of individual galaxies in clusters of galaxies. Astronomers define a galaxy cluster (or cluster of galaxies) as an accumulation of galaxies (from 10 to 100s or even a few thousand members) that lie within a few million light-years of each other and are bound by gravitation. Without the presence of this postulated dark matter, the galaxies in a particular cluster would have drifted apart and escaped from each other's gravitational influence a long time ago. Physicists are engaged now in a variety of research efforts to better define and understand the phenomenon of dark matter.

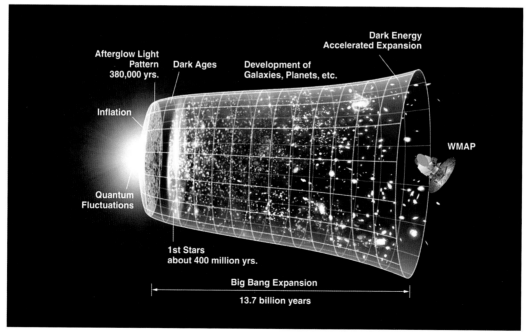

This figure provides a time line of the universe. Events are based on scientific data, including detailed measurements of the cosmic microwave background as performed by NASA's *Wilkinson Microwave Anisotropy Probe (WMAP)* spacecraft. *(NASA/WMAP Science Team)*

The illustration on page 26 depicts the evolution of the universe over the past 13.7 billion years from the big bang event to the present day—as measured by NASA's *Wilkinson Microwave Anisotropy Probe (WMAP)* spacecraft. The far left side of this illustration shows the earliest moment scientists can now investigate—the time when an unusual period of inflation (discussed shortly) produced a burst of exponential growth in the young universe. (NASA illustrators have portrayed size by the vertical extent of the grid.) For the next several billion years, the expansion of the universe gradually slowed down as the matter in the universe (both ordinary and dark) tugged on itself through gravitational attraction. More recently, the expansion of the universe started to speed up again, as the repulsive effects of dark energy became dominant.

The afterglow light seen by *WMAP* was emitted about 380,000 years after the rapid expansion event called inflation and has traversed the universe largely unimpeded since then. When scientists examined the CMB carefully, they discovered telltale signatures imprinted on the afterglow light, revealing information about the conditions of the very early universe. The imprinted information helps scientists understand how primordial events caused later developments (such as the clustering of galaxies) as the universe expanded.

When scientists peer deep into space with their most sensitive instruments, they also look far back in time—eventually viewing the moment when the early universe transitioned from an opaque gas to a transparent gas. Beyond that transition point, any earlier view of the universe remains obscured. The CMB serves as the very distant wall that surrounds and delimits the edges of the observable universe. Scientists can only observe cosmic phenomena back to the remnant glow of this primordial hot gas from the big bang. The afterglow light experienced Doppler shift to longer wavelengths due to the expansion of the universe. The CMB currently resembles emission from a cool dense gas at a temperature of only 4.905 R (2.725 K).

The very subtle variations observed in the CMB are challenging big bang cosmologists. They must explain how clumpy structures of galaxies could have evolved from a previously assumed smooth (that is, uniform and homogeneous) big bang event. Launched on June 30, 2001, NASA's *WMAP* made the first detailed full-sky map of the CMB—including high-resolution data of those subtle CMB fluctuations that represent the primordial seeds that germinated into the cosmic structure scientists observe today. The patterns detected by *WMAP* represent tiny temperature differences within the very evenly dispersed microwave light bathing the universe—a light that now averages a frigid 4.905 R (2.725 K). The subtle CMB fluctuations have encouraged physicists to make modifications in the original big bang hypothesis. One of these modifications involves a concept they call inflation. During inflation, vacuum state fluctuations gave rise to the very rapid (exponential), nonuniform expansion of the early universe.

In 1980, the American physicist Alan Harvey Guth (1947–   ) proposed inflation in order to solve some of the lingering problems in cosmology that were not adequately treated in standard big bang cosmology. As Guth and other later scientists suggested, between 10^{-35} and 10^{-33} second after the big bang, the universe actually expanded at an incredible rate—far in excess of the speed of light. In this very brief period, the universe increased in size by at least a factor of 10^{30}—growing from an infinitesimally small subnuclear-sized dot to a region about 9.84 feet (3 m) across. (Physicists point out that while relativity theory states matter cannot exceed the speed of light, such restrictions are not imposed upon empty space.) By analogy, imagine a grain of very fine sand becoming the size of the presently observable universe in one-billionth (10^{-9}) the time it takes light to cross the nucleus of an atom. During inflation, space itself expanded so rapidly that the distances between points in space increased greater than the speed of light. Scientists suggest that the slight irregularities they now observe in the CMB are evidence (the fossil remnants or faint ghosts) of the quantum fluctuations that occurred as the early universe inflated.

Astrophysicists have three measurable signatures that strongly support the notion that the present universe evolved from a dense, nearly featureless hot gas—just as the big bang theory suggests. These scientifically measurable signatures are the expansion of the universe, the abundance of the primordial light elements (hydrogen, helium, and lithium), and the CMB radiation. Hubble's observation in 1929 that galaxies were generally receding (when viewed from Earth) provided scientists with their first tangible clue that the big bang theory might be correct. The big bang model also suggests that hydrogen, helium, and lithium nuclei should have fused from the very energetic collisions of protons and neutrons in the first few minutes after the big bang. (Scientists call this process big bang nucleosynthesis.) The observable universe has an incredible abundance of hydrogen and helium. Finally, within the big bang model, the early universe should have been very, very hot. Scientists regard the CMB (first detected in 1964) as the remnant heat left over from an ancient fireball.

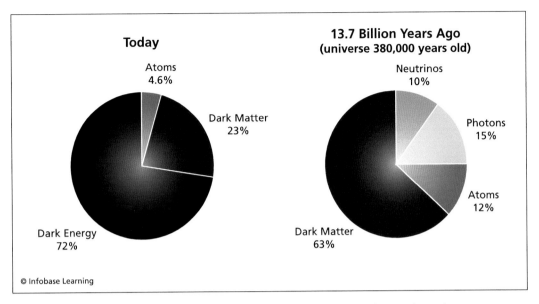

These pie charts describe the content of the past and present universe, based upon evaluation of five years of data from *WMAP*. (Note that *WMAP* data are only accurate to two digits, so the total appearing for today's universe is not exactly 100 percent.) *(NASA/WMAP Science Team)*

The importance of the CMB within modern cosmology cannot be overstated. In March 2008, NASA released the results of a five-year investigation of the oldest light in the universe. Based on a careful evaluation of *WMAP* data, scientists were able to gain incredible insight into the past and present content of the universe. The *WMAP* data (see illustration above) revealed that the current contents of the universe include 4.6 percent atoms—the building blocks of stars, planets, and people. In contrast, dark matter comprises 23 percent of the universe. Finally, 72 percent of the current universe is composed of dark energy, which acts like a type of matter-repulsive, antigravity phenomenon.

Scientists speculate that dark energy is distinct from dark matter and is responsible for the present-day acceleration of universal expansion. The NASA-developed illustration of the universe's contents reflects the current limits inherent in the *WMAP*'s ability to estimate the amount of dark matter and dark energy by observing the CMB. Note that *WMAP* data are accurate to two digits, so the total appearing in the content of today's universe is not exactly 100 percent. Despite these minor limitations, the results are rather startling. The content of the current universe and the

HUNTING DARK MATTER WITH THE
HUBBLE SPACE TELESCOPE

Scientists currently think dark matter exceeds the total amount of ordinary (or baryonic) matter in the universe by a factor of five or more. At present, the most significant scientific fact known about dark matter is that it exerts a detectable gravitational influence within clusters of galaxies.

Using data from NASA's *Hubble Space Telescope (HST)*, astronomers were able to observe how dark matter behaves during a gigantic collision between two galaxy clusters. The colossal cosmic wreck created a ripple of dark matter. Cleverly, scientists generated the map of the suspected dark matter distribution (shown in the illustration on page 31) by making very careful observations of how gravitational lensing affected the light from more distant galaxies. Einstein's general theory of relativity predicted gravitational lensing—an intriguing imaging phenomenon caused by the bending, distortion, and magnification of the light from distant cosmic sources as that light passes through less distant gravitational fields.

The composite image (on page 31) depicts a (scientist-generated) deep blue, ghostly ring of dark matter in the galaxy cluster Cl 0024+17. The ringlike structure of the deep blue map corresponds to the cluster's dark matter distribution. To create this composite image, NASA scientists carefully superimposed the shadowy deep blue colored map on top of a more familiar, visible image of the same galaxy cluster.

early universe (about 380,000 years after the big bang) are quite different. This suggests the persistent influence of the cosmic tug-of-war between energy and matter.

HOW HYDROGEN AND HELIUM FORMED
IN THE ANCIENT INFERNO

The most significant step in the evolution of a universe filled with matter and life occurred when hydrogen and helium formed within three minutes after the big bang. Any discussion about the energy of matter should include this very important milestone.

The discovery of the ring of dark matter in galaxy cluster Cl 0024+17 represents some of the strongest evidence testifying to the existence of dark matter. Since the 1930s, astronomers and astrophysicists have suspected the presence of some gravitationally influential, invisible substance holding clusters of galaxies together. Scientists estimated that the masses associated with the visible stars in such clusters were simply not enough to generate the gravitational forces needed to keep the whirling galactic clusters together.

A *Hubble Space Telescope (HST)* composite image that shows a ghostly, deep blue ring of dark matter superimposed upon a visible *HST* image of galaxy cluster Cl 0024+17. The scientist-generated, deep blue map of the cluster's dark matter distribution is based upon a phenomenon called gravitational lensing. Although invisible, dark matter is inferred present, because its gravitational influence bends the incoming light from more distant galaxies. *(NASA, ESA, M. J. Lee, and H. Ford [Johns Hopkins University])*

Contemporary scientific measurements estimate that hydrogen makes up approximately 73 percent of the mass of ordinary matter in the observable universe. Helium makes up about 25 percent of the mass, and everything else is only 2 percent. While the relative abundances of the chemical elements beyond hydrogen and helium appear quite low in these estimates, it is important to realize that most of the atoms here on Earth, including those in a person's body, are part of this small portion of ordinary matter. The primordial low mass nuclei (hydrogen, deuterium, helium-3, helium-4, and lithium-7 [only trace quantities]) were forged in the crucible of the very early universe just minutes after the big bang. At that time, temperatures were about 1.8×10^9 R (10^9 K) and big bang nucleosynthesis took place. These low mass nuclei owe their existence and elemental heritage to the intensely hot and very dense conditions accompanying the birth of the universe.

Although some of the details still remain a bit sketchy (due to limitations in today's physics), this section explains where the elements hydrogen and helium came from. The next section describes how all the other chemical elements found here on Earth (and elsewhere in the universe) were created by nuclear reactions during the life and death of ancient stars.

Physicists call the incredibly tiny time interval between the big bang event and 10^{-43} second, the Planck time or the Planck era in honor of the German physicist Max Planck (1858–1947), who proposed quantum theory. Current levels of science are inadequate to describe the extreme conditions of the very young universe during the Planck era. Scientists can only speculate that during the Planck era the very young universe was extremely small and extremely hot—perhaps at a temperature greater than 1.8×10^{32} R (10^{32} K). At such an unimaginably high temperature, the four fundamental forces of nature were most likely merged into one unified

RANKINE—THE OTHER ABSOLUTE TEMPERATURE

Most of the world's scientists and engineers use the Kelvin scale (named after the Scottish physicist William Thomson, first baron Kelvin [1824–1907]) to express absolute thermodynamic temperatures. But there is another absolute temperature scale, called the Rankine scale (symbol R), that sometimes appears in engineering analyses performed in the United States—analyses involving American customary units.

In 1859, the Scottish engineer and physicist William John Macquorn Rankine (1820–72) introduced the absolute temperature scale that now carries his name. Absolute zero in the Rankine temperature scale (that is, 0 R) corresponds to −459.67°F. The relationship between temperatures expressed in rankines (R) using the (absolute) Rankine scale and those expressed in the degrees Fahrenheit (°F) using the (relative) Fahrenheit scale is: T (R) = T (°F) + 459.67. The relationship between the Kelvin scale and the Rankine scale is: (9/5) × absolute temperature (kelvins) = absolute temperature (rankines). For example, a temperature of 100 K is expressed as 180 R. The use of absolute temperatures is very important in science.

super force. Limitations imposed by the uncertainty principle of quantum mechanics prevent scientists from developing a model of the universe between the big bang event and the end of Planck time. This shortcoming forged an interesting intellectual alliance between astrophysicists (who study the behavior of matter and energy in the large-scale universe) and high energy nuclear particle physicists (who investigate the behavior of matter and energy in the subatomic world).

In an effort to understand conditions of the very early universe, researchers are trying to link the physics of the very small (as described by quantum mechanics) with the physics of the very large (as described by Einstein's general relativity theory). What scientists hope to develop is a new realm of physics they call quantum gravity. If developed, quantum gravity would be capable of treating Planck era phenomena. An important line of contemporary investigation involves the use of very powerful particle accelerators here on Earth to briefly replicate the intensely energetic conditions found in the early universe. Particle physicists share the results of their very energetic particle-smashing experiments with astrophysicists and cosmologists, who are looking far out to the edges of the observable universe for telltale signatures suggesting how the universe evolved to its present-day state from an initially intense inferno.

Particle physicists currently postulate that the early universe experienced a specific sequence of phenomena and phases following the big bang explosion. They generally refer to this sequence as the standard cosmological model. Right after the big bang, the present universe was at an incredibly high temperature. Physicists suggest the temperature was at the unimaginable value of about 1.8×10^{32} R (10^{32} K). During this period, which they sometimes call the quantum gravity epoch, the force of gravity, the strong force, and a composite electroweak force all behaved as a single unified force. Scientists further postulate that at this time the physics of elementary particles and the physics of space-time were one and the same. In an effort to adequately describe quantum gravity, many physicists are now searching for a theory of everything (TOE).

At Planck time (about 10^{-43} second after the big bang), the force of gravity assumed its own identity. With a temperature estimated to be about 1.8×10^{32} R (10^{32} K), the entire spatial extent of the universe at that moment was less than the size of a proton.

Scientists call the ensuing period the grand unified epoch. While the force of gravity functioned independently during this period, the strong force and the composite electroweak force remained together and continued to act like a single force. Today, physicists apply various grand unified theories (GUTs) in their efforts to model and explain how the strong force and the electroweak force functioned as one during this particular period in the early universe.

About 10^{-35} second after the big bang, the strong force separated from the electroweak force. By this time, the expanding universe had cooled to about 1.8×10^{28} R (10^{28} K). Physicists call the period between about 10^{-35} s and 10^{-10} s the electroweak force epoch. During this epoch, the weak force and the electromagnetic force became separate entities, as the composite electroweak force disappeared. From that time forward, the universe contained four fundamental natural forces—namely, the force of gravity, the strong force, the weak force, and the electromagnetic force.

FUNDAMENTAL FORCES IN NATURE

At present, physicists recognize the existence of four fundamental forces in nature: gravity, electromagnetism, the *strong force,* and the *weak force.* Gravity and electromagnetism are part of a person's daily experiences. These two forces have an infinite range, which means they exert their influence over great distances. The two stars in a binary star system, for example, experience mutual gravitational attraction even though they may be a light-year or more distant from each other.

The other two forces, the strong force and the weak force, operate within the realm of the atomic nucleus and involve elementary particles. These forces lie beyond a person's normal experiences and remained essentially unknown to the physicists of the 19th century despite their good understanding of classical physics, including the universal law of gravitation and the fundamental principles of electromagnetism. The strong force operates at a range of about 3.28×10^{-15} feet (1×10^{-15} m) and holds the atomic nucleus together. The weak force has a range of about 3.28×10^{-17} feet (1×10^{-17} m) and is responsible for processes like beta decay that tear nuclei and elementary particles apart. What is important to recognize here is that whenever anything happens in the universe—that is, whenever an object experiences a change in motion—the event takes place because one or more of these fundamental forces is involved.

Following the big bang and lasting up to about 10^{-35} s, there was no distinction between quarks and leptons. All the minute particles of matter were similar. However, during the electroweak epoch, quarks and leptons became distinguishable. This transition allowed quarks and antiquarks to eventually become hadrons, such as neutrons and protons, as well as their antiparticles. At 10^{-4} s after the big bang, in a period scientists call the radiation-dominated era, the temperature of the universe cooled to 1.8 $\times$ 10^{12} R (10^{12} K). By that time, most of the hadrons had disappeared due to matter-antimatter annihilations. The surviving protons and neutrons represented only a small fraction of the total number of particles in the early universe—the majority of which consisted of leptons, such as electrons, positrons, and neutrinos. But, like most of the hadrons before them, most of the leptons also quickly disappeared because of matter-antimatter interactions.

At the beginning of the period scientists call the matter-dominated era, which occurred some three minutes (or 180 seconds) following the big bang, the expanding universe had cooled to a temperature of 1.8 $\times$ 10^9 R (10^9 K) and low-mass nuclei, such as deuterium, helium-3, helium-4, and (a trace quantity of) lithium-7, began to form due to very energetic collisions in which the reacting particles joined or fused. Once the temperature of the expanding universe dropped below 1.8 $\times$ 10^9 R (10^9 K), environmental conditions no longer favored the fusion of low-mass nuclei and the process of big bang nucleosynthesis ceased. Physicists point out that the observed amount of hydrogen and helium in the universe agrees with the theoretical predictions concerning big bang nucleosynthesis.

Finally, when the expanding universe reached an age of about 380,000 years, the temperature had dropped to about 5,400 R (3,000 K), allowing electrons and nuclei to combine and form hydrogen and helium atoms. Within this period of atom formation, the ancient fireball became transparent and continued to cool from a hot 5,400 R (3,000 K) down to a frigid 4.905 R (2.725 K)—the currently measured temperature of the CMB.

COSMIC ORIGIN OF THE CHEMICAL ELEMENTS

This section provides a brief summary of the origin of the chemical elements beyond hydrogen and helium. The cosmic sources include energetic processes within stellar fusion furnaces and gigantic end-of-life explosions called supernovas.

There were not any stars shining when the universe became transparent some 380,000 years after the big bang. Thus, scientists refer to the ensuing period as the cosmic dark ages. During the cosmic dark ages, very subtle variations (inhomogeneities) in density allowed the attractive force of gravity to form great clouds of hydrogen and helium. Gravity continued to subdivide these gigantic gas clouds into the smaller clumps, enabling the formation of the first stars. As many brilliant new stars appeared, they gathered into galaxies. Astronomers currently estimate that about 400,000 million years after the big bang the cosmic environmental conditions were right for a rapid rate of star formation. Soon after, the cosmic dark ages became illuminated by the radiant emissions of millions of young stars (called population III stars).

Gravitational attraction slowly gathered clumps of hydrogen and helium gas into new stars. The very high temperatures encountered in the cores of massive early stars supported the manufacture of heavier nuclei up to and including iron by means of a process scientists call nucleosynthesis. Elements beyond iron were formed in a bit more spectacular fashion. Neutron capture processes deep inside highly evolved massive stars and subsequent supernova explosions at the end of their relatively short lifetimes synthesized all the elements beyond iron.

A star forms when a giant cloud of mostly hydrogen gas, perhaps light-years across, begins to contract under its own gravity. This clump of hydrogen and helium gas eventually collects into a giant ball that is hundreds of thousands of times more massive than Earth. As the giant gas ball continues to contract under its own gravitational influence, an enormous pressure arises in its interior. The increase in pressure at the center of the protostar is accompanied by an increase in temperature. When the center of the contracting gas ball reaches a minimum temperature of about 18 million rankines (10 million kelvins), the hydrogen nuclei acquire sufficient velocities to experience collisions that support nuclear fusion reactions. (The central temperature of the Sun's core is approximately 27×10^6 R [15×10^6 K].) This is the moment when a new star is born. The process of nuclear fusion releases a great quantity of energy at the center of the young star. Once thermonuclear burning begins in the star's core, the energy released counteracts the continued contraction of stellar mass by gravity. As the inward pull of gravity exactly balances the outward radiant pressure from thermonuclear fusion reactions in the core

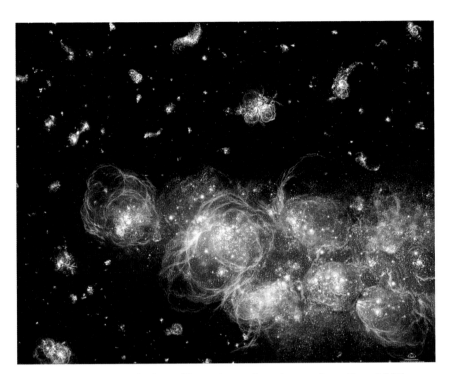

This is an artist's rendering of how the early universe (less than 1 billion years after the big bang) must have looked when it went through a very rapid onset of star formation, converting primordial hydrogen into a myriad of stars at an unprecedented rate. Regions filled with newborn stars began to glow intensely. Brilliant new stars began to illuminate the cosmic dark ages. The most massive of these early stars self-detonated as supernovas, creating and spreading chemical elements throughout the fledgling universe. Analysis of *Hubble Space Telescope* data supports the hypothesis that the universe's first stars (called population III stars) appeared in an eruption of star formation, rather than at a gradual pace. *(NASA, K. Lanzetta [SUNY], and Adolf Schaller [STScI])*

the contracting ball of gas becomes stable. Ultimately, the energy released in fusion flows upward to the star's outer surface, and the new star radiates its energy into space.

Size definitely matters in stellar astronomy. Stars come in a variety of sizes, ranging from about one-tenth to 60 (or more) times the mass of the Sun. It was not until the mid-1930s that scientists began to recognize that the process of nuclear fusion takes place in the interiors of all normal stars and fuels their enormous radiant energy outputs. Scientists use the term *nucleosynthesis* to describe the complex process of how different size stars create different elements through nuclear fusion reactions and various alpha particle (helium nucleus) capture reactions. All stars on the main sequence use thermonuclear reactions to convert hydrogen into helium, liberating energy in the process. The initial mass of a star determines not only how long it lives (as a main sequence star) but also how it dies.

Small- to average-mass stars share the same fate at the end of their relatively long lifetimes. At birth, low-mass stars begin their stellar lives by fusing hydrogen into helium within their cores. This process generally continues for billions of years, until there is no longer enough hydrogen in a particular star's core to fuse into helium. Once hydrogen burning stops, so does the release of the thermonuclear energy that produced the outward radiant pressure, which counteracted the relentless inward attraction of gravity.

At this point in its lifetime, the small star begins to collapse inward. Gravitational contraction causes an increase in temperature and pressure. Any hydrogen remaining in the star's middle layers soon becomes hot enough to undergo thermonuclear fusion into helium in a shell around the dying star's core. The release of fusion energy in this shell enlarges the star's outer layers, causing the star to expand far beyond its previous dimensions. The expansion process cools the outer layers of the star, transforming them from brilliant white hot or bright yellow in color to a shade of dull glowing red. Astronomers call a low-mass star at this point in its lifetime cycle a red giant.

Gravitational attraction continues to make the low-mass star collapse until the pressure in its core reaches a temperature of about 180×10^6 R (100×10^6 K). This very high temperature is sufficient to allow the thermonuclear fusion of helium into carbon. The fusion of helium into carbon now releases enough energy to prevent further gravitational collapse—at least until the helium runs out. This stepwise process continues until oxygen is fused. When there is no more material to fuse at the progressively increasing high temperature conditions within the collapsing core, gravity again exerts it relentless attractive influence on matter. This time, the heat released during gravitational collapse causes the outer layers of the low-mass star to blow off, creating an expanding symmetrical cloud of material that astronomers call a planetary nebula. The planetary nebula may contain up to 10 percent of the dying star's mass. The explosive blow-off process is very important, because it disperses into space some of the low-mass elements (like carbon) that were manufactured by the small dying star.

The final collapse that causes the small star to eject a planetary nebula also releases energy. But this energy release is insufficient to fuse other elements. The remaining core material continues to collapse until all the atoms are crushed together and only the repulsive force between the electrons counteracts gravity's pull. Physicists refer to this type of very condensed matter as degenerate matter. The final compact object is a degenerate matter star called a white dwarf. The white dwarf represents the final phase in the evolution of most low-mass stars, including the Sun.

In some very rare cases, a white dwarf can undergo a gigantic explosion that astrophysicists call a Type Ia supernova. This happens when the white dwarf is part of a binary star system and pulls too much matter from its stellar companion. The moment the compact white dwarf can no longer support the addition of incoming mass from its stellar companion it experiences a rapid, new wave of gravitational collapse, followed by runaway thermonuclear burning, and completely explodes. Nothing is left behind. All the chemical elements created during the lifetime of the low-mass star are now scattered into space in a spectacular detonation.

Astrophysicists define large stars as those with more than three to five times the mass of the Sun. Large stars begin their lives in pretty much the same way as small stars by fusing hydrogen into helium. Because of their size, large stars burn their thermonuclear fuel faster and hotter—typically fusing all the hydrogen in their cores into helium in less than 1 billion years. Once the hydrogen in a large star's core is fused into helium, it becomes a red supergiant—a stellar object similar to a red giant, only much larger. The red supergiant star experiences very high core temperatures during gravitational contraction and begins to fuse helium into carbon, carbon and helium into oxygen, and even two carbon nuclei into magnesium. Thus, through a combination of intricate nucleosynthesis reactions, the red supergiant forms progressively heavier elements up to and including the element iron. Astrophysicists suggest that the red supergiant forms an onionlike structure—with different elements being fused at different temperatures in layers around the core.

When nuclear reactions begin to fill the core of a red supergiant with iron, the overall thermonuclear energy release in the interior begins to decrease. (Iron nuclei do not undergo energy-liberating nuclear fusion reactions.) Because of this decline, the dying massive star no longer has the internal radiant pressure to resist the attractive force of gravity. The sudden gravitational collapse causes the core temperature to rise to more than 180×10^9 R (100×10^9 K), smashing the electrons and protons in each iron atom together to form neutrons. The force of gravity now draws this massive collection of neutrons incredibly close together. For about a second, the neutrons fall very fast toward the highly compressed center of

This mosaic image is one of the largest ever taken by NASA's *Hubble Space Telescope* of the Crab Nebula in the constellation of Taurus. The image shows the expanding remnant of a supernova explosion that is about six light-years wide. The orange filaments are the tattered remains of the star and consist mostly of hydrogen. A rapidly spinning neutron star (the crushed, ultradense core of the exploded star) is embedded in the center of the nebula and serves as the dynamo powering the nebula's eerie interior bluish glow. Chinese astronomers witnessed this violent event in 1054 C.E. *(NASA, ESA, J. Hester [Arizona State University])*

the star. Then, while they are crunching each other, the neutrons abruptly stop. This sudden stop causes the neutrons to recoil violently and an explosive shock wave travels outward from the highly compressed core. As the shock wave travels from the core, it heats and accelerates the outer layers of material of the red supergiant, causing the majority of the large star's mass to be blown off into space. Astrophysicists call this enormous explosion a Type II supernova. The stellar core end product of a Type II supernova is either a neutron star or a black hole.

Astrophysicists suggest that the slow neutron capture process (or s-process) produces heavy nuclei up to and including bismuth-209, the most massive, naturally occurring, nonradioactive nucleus. The flood of neutrons that accompanies a Type II supernova explosion is responsible for the rapid neutron capture process (or r-process). It is these rapid neutron capture reactions that form the radioactive nuclei of the heaviest elements typically found in nature, such as thorium-232 and uranium-238. The violently explosive force of the supernova also hurls stellar-minted elements throughout interstellar space in a dramatic shower of stardust. The expelled stardust eventually combines with interstellar gas. The elementally enriched interstellar gas then becomes available to create a new generation of stars—including those with a family of planets.

Population I stars are hot, luminous young stars (like the Sun) that reside in the disk of a spiral galaxy and are higher in heavy element content (about 2 percent abundance) than population II stars. Population II stars are older stars that are generally found in globular clusters as well as in the halo of a galaxy. The halo is the distant spherical region that surrounds a galaxy.

About 4.6 billion years ago, the solar system (including planet Earth) formed from an elementally enriched primordial cloud of hydrogen and helium gas. All things animate and inanimate here are on Earth are the natural by-products of the life and death of massive ancient stars.

CHAPTER 3

The Energy of Motion

If I have seen further than others, it is by standing on the shoulders of giants.

—Sir Isaac Newton (1676)

As part of the Scientific Revolution, the combined achievements of Galileo Galilei (1564–1642), Johannes Kepler (1571–1630), and Sir Isaac Newton (1642–1727) provided the first useful quantitative interpretation of matter in motion. This chapter discusses the origin of the classical interpretations of kinetic energy, potential energy, and work.

THE INFLUENCE OF GRAVITY

Gravity is the attractive force that tugs on people and holds them on the surface of Earth. Gravity also keeps the planets in orbit around the Sun and causes the formation of stars, planets, and galaxies.

Where there is matter, there is gravity. In a simple definition, gravity is the pull material objects exert on each other. During a football game, when a quarterback attempts to complete a long pass to a receiver far downfield, the football travels into the air, rises to a certain height, pauses briefly at the apex of its trajectory, and then spirals to Earth under the influence of gravity. If all goes as planned, the receiver catches the ball before it hits the ground or gets deflected by a player on the opposing team.

When artillery troops fire their cannon, the projectile flies out of the barrel, travels through the air at high velocity, and hits on or near its intended target. Once the projectile of a traditional artillery shell leaves the cannon's barrel, scientists consider it a high-speed, free-flying projectile that follows a ballistic trajectory under the influence of gravity and aerodynamic forces.

Finally, some of the most popular rides in amusement parks around the world are roller coasters. As riders rapidly travel around specially constructed closed pathways, these electromechanical systems entertain thrill-seeking patrons by continuously transforming potential energy and kinetic energy.

Today, most people recognize the attractive influence gravity exerts on material objects whether here on Earth or elsewhere in the universe. Yet just 400 years ago, at the start of the 17th century, no one had an inkling of the connection between the motion of celestial bodies and the behavior of objects falling to the ground here on Earth. Two very different but determined individuals—the Italian scientist Galileo Galilei and the German astronomer Johannes Kepler—made important contributions to understanding the energetic behavior of moving matter. Their pioneering scientific activities helped change the course of human history. The British scientist Sir Isaac Newton built upon their work and developed his insightful, quantitative interpretations of mass, motion, acceleration,

A wide receiver for the Air Force Academy's football team catches a pass during an intercollegiate game at Falcon Stadium, Colorado Springs, in 2009. *(U.S. Air Force)*

A 155-mm artillery shell hurtles out of the barrel of an M-198 howitzer during live fire and maneuver training by U.S. Marines at the Al Hamra Training Area in the United Arab Emirates in November 2000. *(DOD/USMC)*

and force. Newton's model of the mechanical universe still dominates classical physics. In 1676, Newton eloquently acknowledged the contributions of Galileo and Kepler, when he wrote: "If I have seen further than others, it is by standing on the shoulders of giants."

GALILEO GALILEI

Galileo Galilei is often remembered as the first astronomer to use a telescope to view the heavens and to conduct early astronomical observations that promoted the Scientific Revolution. But he was also the physicist who founded the science of mechanics and provided Newton the underlying data and ideas that allowed the British physicist to develop the laws of motion and the universal law of gravitation.

Galileo Galilei was born in Pisa on February 15, 1564. (Scientists commonly refer to Galileo by his first name only.) In 1585, he left university without receiving a degree and focused his activities on the physics of solid bodies. The motion of falling objects and projectiles intrigued him. In 1589, Galileo became a mathematics professor at the University of Pisa. He was a brilliant lecturer and students came from all over Europe to attend his classes. This circumstance quickly angered many senior, but less capable, faculty members. To make matters worse, Galileo often used his tenacity, sharp wit, and biting sarcasm to win philosophical arguments at the university. His tenacious and argumentative personality earned him the nickname the "Wrangler."

In the late 16th century, European professors usually taught natural philosophy (physics) as metaphysics—an extension of Aristotelian philosophy. Before Galileo's pioneering contributions, physics was not

This 1983 Italian stamp honors Galileo Galilei, the brilliant physicist and astronomer, whose pioneering activities promoted the Scientific Revolution. *(courtesy of the author)*

regarded as an observational or experimental science. But through his skillful use of mathematics and innovative experiments, Galileo changed that approach and established the important process known as the scientific method. Galileo's research activities constantly challenged the 2,000-year tradition of ancient Greek learning.

Aristotle had stated that heavy objects would fall faster than lighter objects. Galileo disagreed and held the opposite view that, except for air resistance, the two objects would fall at the same time regardless of their masses. It is not certain whether he personally performed the legendary musket ball versus cannonball drop experiment from the Leaning Tower of Pisa to prove this point. He did, however, conduct a sufficient number of experiments with objects rolling or sliding down inclined planes to upset Aristotelian thinking and create the science of mechanics.

During his lifetime, Galileo was limited in his motion experiments by an inability to accurately measure small increments of time. No one had yet developed a timekeeping device capable of accurately measuring 10ths, 100ths, or 1,000ths of a second. Despite this severe impediment, he conducted many important experiments that produced remarkable insights into the physics of free fall and projectile motion. Newton would build upon Galileo's pioneering work to create the universal law of gravitation and the three laws of motion—intellectual pillars of classical physics.

Galileo's experiments demonstrated that in the absence of air resistance, all bodies, regardless of mass, fall to the surface of Earth at the same acceleration. Acceleration is the rate at which the velocity of an object changes with time. On or near Earth's surface, the acceleration due to gravity (symbol: g) of a free-falling object has the standard value of 32.1737 ft/s^2 (9.80665 m/s^2) by international agreement. Galileo's work in solid mechanics, including the physics of projectile motion and free-falling bodies, led to the incredibly important formula that helped Newton create his laws of motion and universal law of gravitation. The formula ($d = \frac{1}{2} g t^2$) mathematically described the distance (d) traveled by an object in free fall (neglecting air resistance) in time (t). An interesting example follows: How long would it take a dense solid object, like a baseball, to reach the ground—if a person drops it (carefully) from the Leaning Tower of Pisa? The height of this famous bell tower measures about 184 ft (56 m) from the lowest side to the ground. Neglecting the influence of air resistance, an object dropped from the lowest side of the Leaning Tower would take approximately 3.38 seconds to hit the ground.

During NASA's *Apollo 15* landing mission to the Mare Imbrium (Sea of Rains) region of the Moon, the American astronaut David R. Scott (1932–   ) performed a televised physics demonstration (on August 2, 1971), in which he imitated the legendary cannonball drop from the Leaning Tower of Pisa. In this extraterrestrial science demonstration, Scott simultaneously released a feather and a hammer, which in the lunar vacuum fell to the Moon's surface at the same rate. Assuming he released the two objects from a height of 4.9 feet (1.5 m) above the lunar surface, it took 1.37 seconds for the geologist's hammer and falcon's feather to simultaneously hit the surface. Remember, the local acceleration of gravity on the surface of the Moon is just 5.25 ft/s^2 (1.6 m/s^2), because the Moon is much smaller and less massive than Earth.

(opposite) According to popular scientific legend, Galileo dropped a cannonball and a musket ball from the Leaning Tower of Pisa to demonstrate that (save for air resistance) objects fall at the same rate under the influence of gravity. While Galileo never documented this particular experiment in his notes, he did perform a large number of experiments with objects rolling or sliding down inclined planes. Through these important experiments, Galileo upset Aristotelian "physics" and created the modern science of mechanics. *(U.S. Air Force)*

Scott's simple demonstration validated the universality of the physical laws that emerged from the great intellectual accomplishments of Galileo, Kepler, Newton, and other scientists, who looked up at the Moon, planets, and stars and wondered. One of the most important contributions of Western civilization to the human race is the emergence of modern science and the development of the scientific method. During the intellectually turbulent period of the 17th century, men of great genius, such as Galileo, discovered important physical laws and constructed experimental tech-

FREE-FALLING BODIES

Scientists define free fall as the unimpeded fall of an object in a gravitational field. The atmospheric plunge of a skydiver before his parachute has opened is an example of free fall. Physicists often idealize the motion of objects in free fall near Earth's surface by neglecting the influence of air resistance and assuming the acceleration of gravity *(g)* has an average constant value of about 32.2 ft/s² (9.8 m/s²).

The acceleration due to gravity is directed downward toward the center of Earth. In reality, the value of this acceleration decreases with altitude above Earth's surface and even varies slightly with respect to latitude. The work of Galileo, Newton, and other scientists demonstrated that, *neglecting air resistance,* an object released from a distance (d) above Earth's surface, would experience free-fall behavior according to the following formula: $d = \frac{1}{2} g t^2$, where d is the vertical distance (height) in feet (ft) or meters (m), *g* is the acceleration of gravity (ft/s² [m/s²]), and t is time (in seconds [s]). This deceptively simple relationship supported the establishment of solid-body mechanics. It was Galileo's detailed investigation of the physics of free fall that helped Newton unlock the secrets of motion of the mechanical universe.

When an object drops from a height above Earth's surface, the planet's gravitational attraction causes the object's velocity to increase. After a few seconds of free fall, the object's velocity is substantial and air resistance becomes significant. Eventually, the resistive force of the atmosphere balances the attractive force of gravity and the object no longer accelerates, as it plunges toward Earth. At this point in its free-fall descent, scientists say the object has reached terminal velocity. From that moment on during free fall, the object continues its downward flight at a constant (terminal) velocity without experiencing any

niques that helped people explain how matter and energy influenced the operation and behavior of the physical universe.

UNLOCKING THE SECRETS OF PLANETARY MOTION

The German astronomer and mathematician Johannes Kepler was a contemporary of Galileo and also helped spread the Scientific Revolution.

further acceleration. Many factors determine the terminal velocity of a free-falling object, including its mass, aerodynamic shape, and overall surface area. A skydiving parachutist, who delays opening his chute after jumping from an airplane, will typically achieve a terminal velocity of about 184 ft/s (56 m/s) or about 124 MPH (200 km/hr).

Pararescue specialists stationed at Kadena Air Base conduct a free-fall jump out of a MC-130P aircraft at 10,000 feet (3,048 m) during a training mission over le Shima Island, Japan, in January 2009. *(U.S. Air Force)*

While Galileo's pioneering use of the telescope provided the observational foundations for Copernican cosmology, Kepler's innovative mathematical contributions to astronomy established its theoretical foundations. Through painstaking and arduous analyses, Kepler developed his three laws of planetary motion—important physical principles that describe the elliptical of orbits of planets around the Sun. His work not only provided the empirical basis for the early acceptance of the heliocentric hypothesis of the Polish astronomer Nicholas Copernicus (1473–1543), but it also provided the mathematical starting point from which Newton would develop his law of gravitation.

He was born on December 27, 1571, in the small town of Weil der Stadt, Württemberg, Germany (then part of the Holy Roman Empire). His father was a mercenary soldier and his mother was the daughter of an innkeeper. A sickly child, Kepler suffered from smallpox at the age of three. The disease impaired the use of his hands and affected his eyesight for the remainder of his life. When Kepler was five, his father left to fight in one of the many local conflicts then ravaging Europe. He never returned home. So Kepler spent the remainder of his childhood living with his mother at his grandfather's inn. During his childhood, his mother took him outside the walls of the city so he could witness the great comet of 1577. This special event appears to have encouraged his lifelong interest in astronomy.

After completing local schooling, he pursued a religious education at the University of Tübingen in the hopes of becoming a Lutheran minister. He graduated in 1588 with his baccalaureate degree and then went on to complete a master's degree in 1591. While Kepler was a student at Tübingen, his professor of astronomy, Michael Maestlin (1550–1632), introduced him to the Copernican hypothesis. Kepler immediately embraced the new heliocentric model of the solar system. As his interest in the motion of the planets grew, Kepler's skill in mathematics also emerged. By 1594, partially because he alienated church officials after expressing his personal disagreement with certain aspects of Lutheran doctrine, he totally abandoned any plans for the Lutheran ministry and became a mathematics instructor in Graz, which is a city located in modern Austria.

Despite being an accomplished astronomer and mathematician, Kepler also maintained a strong, lifelong interest in mysticism. He extracted many of his mystical notions from the early Greeks, including the music of the celestial spheres that was originally suggested by Pythagoras in the sixth century B.C.E. As was common for many 17th-century astronomers,

Kepler also dabbled in astrology. He often cast horoscopes for important benefactors, such as Emperor Rudolf II (1552–1612) and Duke (Imperial General) Albrecht von Wallenstein (1583–1632). Astrology provided the frequently impoverished Kepler with a supplemental source of income. His well-prepared horoscopes earned him the political protection of high-ranking officials in the Holy Roman Empire. He was often the only prominent Protestant allowed to live in a German city under control of a Catholic ruler.

In 1596, Kepler published *Mysterium Cosmographicum* (Cosmographic mystery)—an intriguing work in which he unsuccessfully tried to analytically relate Plato's five basic geometric solids (as found in early Greek philosophy and mathematics) to the distances of the six known planets from the Sun. This interesting work, often considered the first outspoken defense of the Copernican hypothesis, attracted the attention of Tycho Brahe (1546–1601), the most famous (pre-telescope) astronomer of the period.

Kepler married his first wife in 1597. She was a wealthy widow named Barbara Mueller. When all the Protestants were forced to leave the Catholic-controlled city of Graz in 1598, Kepler found it extremely difficult to liquidate her property holdings. Any nuptial financial comfort quickly dissipated as the Kepler family hastily departed Graz and moved to Prague in response to Tycho Brahe's invitation.

Kepler joined the elderly Danish astronomer in 1600 as his assistant. When Brahe died the following year, Kepler succeeded him as the imperial mathematician to the Holy Roman Emperor Rudolf II. As a result, Kepler acquired all the precise astronomical data Brahe had collected. These data, especially Brahe's precise observations of the motion of Mars, played an important role in the development of Kepler's three laws of planetary motion.

In 1604, Kepler wrote the book, *De Stella Nova* (New Star), in which he described a bright new star (now known as a supernova) in the constellation Ophiuchus. He first observed the event on October 9, 1604. Modern astronomers sometimes refer to this supernova event as Kepler's Star. According to astronomical observations in Europe and Korea, this particular supernova remained visible for approximately one year.

From 1604 until 1609, Kepler's main interest was a detailed study of the orbital motion of Mars. Before his death, Brahe had challenged his young assistant with the task of explaining the puzzling motion of the Red Planet. Even accepting the Copernican hypothesis, which Kepler did (but Brahe did not), a circular orbit around the Sun did not properly fit the Danish astronomer's carefully observed orbital position data. With youthful optimism, Kepler told the older astronomer he would have an answer for him in a week. It took Kepler eight years to finally obtain the solution. The movement of Mars could not be explained and predicted unless Kepler assumed the orbit was an ellipse with the Sun located at one focus. This revolutionary assumption produced a major advance in solar system astronomy and provided the first empirical evidence of the validity of the Copernican model.

Kepler recognized that the other observable planets also followed elliptical orbits around the Sun. He published this important discovery in 1609 in the book, *Astronomia Nova* (New astronomy). The book, dedicated to Emperor Rudolf II, confirmed the Copernican model and permanently shattered 2,000 years of geocentric Greek astronomy. Kepler became the first scientist to present a well-written demonstration of the scientific method. This work clearly acknowledged the errors and imperfections in the observational data and then compensated for these inaccuracies by creating a new scientific law (model) that used mathematics to accurately predict the natural phenomenon of interest (here the orbital position of Mars). In this important document, Kepler announced that the orbits of the planets were ellipses with the Sun as a common focus. Today, astronomers call this statement Kepler's first law of planetary motion. Kepler also introduced his second law of planetary motion in this book.

The year 1611 proved extremely difficult for Kepler. His first wife and their seven-year-old son died. Then, his royal patron, Emperor Rudolf II, abdicated the throne in favor of his brother Matthias. Unlike Rudolf, Emperor Matthias did not believe in tolerance for the Protestants living in his realm. So Kepler, along with many other Lutherans, left Prague to avoid the start of an impending civil war between Catholics and Protestants. He moved his surviving children to Linz, where he accepted a position as district mathematician and teacher.

In 1613, desperately needing someone to care for his children, he married Susanna Reuttinger, herself an orphan. Although his second marriage was a generally happy one, Kepler continued to suffer misfortune. He had chronic financial troubles, experienced the deaths of two infant daughters, and had to return to Württemberg to defend his shrewish mother, Katharina Kepler, who was put on trial as a witch. This trial dragged on for three years. Kepler used all his legal wit and expended all his political capital to get her released. Despite these personal stresses, he remained a diligent mathematician in Linz until 1628 and used any available time to write several important books. In 1628, he moved his family to Sagan (in Silesia) in order to serve General Albrecht von Wallenstein as his court mathematician. With the Thirty Years' War (1618–48) raging in Germany, he had to flee Sagan in 1630 to avoid religious persecution when Wallenstein fell from power.

Kepler continued his work involving the orbital dynamics of the planets when he published *De Harmonica Mundi* (Concerning the harmonies of the world) in 1619. Although this book extensively reflected Kepler's fascination with mysticism, it also provided a very significant insight that connected the mean distances of the planets from the Sun with their orbital periods. This discovery became known as Kepler's third law of planetary motion.

Between 1618 and 1621, despite constant relocations, Kepler summarized all his planetary studies in an important seven-volume effort entitled *Epitome Astronomica Copernicanae* (Epitome of Copernican astronomy). This work presents all of Kepler's heliocentric astronomy in a systematic way. As a point of scientific history, Kepler actually based his second law (the law of equal areas) on the mistaken (but reasonable for the time) physical assumption that the Sun exerted a strong magnetic influence on all the planets. Later in the 17th century, Newton provided the right physical explanation when he developed the universal law of gravitation to identify the force that causes the planetary motion so correctly described by Kepler's second law.

Kepler's three laws of planetary motion are: first (the law of ellipses), the planets move in elliptical orbits with the Sun as a common focus; second (the law of areas), as a planet orbits the Sun, the radial line joining the planet to the Sun sweeps out equal areas within the ellipse in equal times; and third (the harmonic law), the square of the orbital period (P) of a planet is proportional to the cube of its mean distance (a) from the Sun. The third law states that there is a fixed ratio between the time it takes a planet to complete an orbit around the Sun and the size of the orbit. Astronomers often express this ratio as P^2/a^3, where a is the semimajor axis of the ellipse and P is the period of revolution around the Sun. Kepler's three laws placed modern (telescopic era) observational astronomy on a solid, mathematical foundation.

Treating the Moon as a satellite of Earth, perigee is the point in the Moon's orbit when it is closest to Earth. At perigee, the Moon has the maximum amount of orbital velocity (kinetic energy) and the minimum amount of gravitational (potential) energy. The Moon is at its greatest distance from Earth at apogee. At apogee, the Moon also has the lowest amount of orbital velocity (kinetic energy) and the highest amount of gravitational (potential) energy. Similarly, at perihelion, Earth is closest to the Sun and has the most orbital velocity; while at aphelion, Earth is farthest from the Sun, has the lowest orbital velocity (kinetic energy), and the highest amount of gravitational potential energy.

In 1627, Kepler's *Tabulae Rudolphinae* (Rudolphine tables), named after his benefactor Emperor Rudolf and dedicated to Tycho Brahe, provided astronomers with detailed planetary position data. The tables remained in use until the 18th century. As a skilled mathematician, Kepler was able to use the logarithm to perform many tedious calculations. This effort was the first important scientific application of the logarithm—a new mathematical function invented by the Scottish mathematician John Napier (1550–1617).

Kepler also made important contributions in the field of optics. While in Prague, he wrote *Optica* (Optics) in 1604. Prior to 1610, Galileo and Kepler communicated with each other, although they never met. According to one historical account of their relationship, in 1610 Kepler refused to believe that Jupiter had four moons that behaved like a miniature solar system unless he personally observed them. When a Galilean telescope arrived at his doorstep, Kepler promptly used the device. Kepler immediately called the four major moons discovered by Galileo satellites, which he derived from the Latin word *satelles* meaning the people who escort or loiter around a powerful person. In 1611, Kepler improved the design of Galileo's original telescope by introducing two convex lenses in place of the one convex lens and one concave lens arrangement used by the great Italian astronomer. Kepler's final published work in Prague was his *Dioptrice* (Dioptics), which he completed there in 1611. This book is regarded as the first scientific work in geometrical optics.

Before his death in 1630, Kepler wrote a very interesting novel called *Somnium* (The dream). It was a story about an Icelandic astronomer who travels to the Moon. While the tale involves demons and witches, who help get the hero to the Moon's surface in a dream state, Kepler's description of the lunar surface was quite accurate. This story, published after his death in 1634, is often regarded by science historians as the first genuine piece of science fiction.

Kepler died on November 15, 1630, in Regensburg, Bavaria, of a fever contracted while journeying to see the emperor. Kepler was trying to collect the payment owed him for his service as imperial mathematician and for his effort in producing *Tabulae Rudolphinae*.

NEWTON'S MECHANICAL UNIVERSE

Sir Isaac Newton was the brilliant though introverted British physicist, mathematician, and astronomer, whose law of gravitation, three laws of motion, development of the calculus, and design of a new type of reflecting telescope made him one of the greatest scientific minds in human history. Through the patient encouragement and financial support of the British mathematician Sir Edmund Halley (1656–1742), Newton published his great work, *Philosophiae Naturalis Principia Mathematica* in 1687. This monumental book transformed the practice of physical science and completed the Scientific Revolution started by Copernicus a century earlier. Newton's three laws of motion and universal law of gravitation still serve as the basis of classical mechanics.

Newton was born prematurely in Woolsthorpe, Lincolnshire, on December 25, 1642 (using the former Julian calendar). His father had died before Newton's birth, and this event contributed to a very unhappy childhood. In 1665, he graduated with a bachelor's degree from Cambridge University without any particular honors or distinction.

Following graduation, Newton returned to the family farm to avoid the plague, which had broken out in London. For the next two years, he pondered mathematics and physics at home, and this self-imposed exile laid the foundation for his brilliant contributions to science. By Newton's own account, one day on the farm he saw an apple fall to the ground and began to wonder if the same force that pulled on the apple also kept the Moon in its place. At the time, heliocentric cosmology as expressed in the works of Copernicus, Galileo, and Kepler was becoming more widely accepted (except where banned for political or religious reasons), but the mechanism for planetary motion around the Sun remained unexplained.

By 1667, the plague epidemic subsided, and Newton returned to Cambridge as a minor fellow at Trinity

This 1993 German stamp honors the British scientist Sir Isaac Newton. *(Department of Energy)*

College. The following year, he received his master's degree and became a senior fellow. Around 1668, he constructed the first working reflecting telescope, an important new astronomical instrument. This telescope design earned Newton a great deal of professional acclaim, including eventual membership in the Royal Society.

In 1669, Isaac Barrow, Newton's former mathematics professor, resigned his position so that the young Newton could succeed him as Lucasian Professor of Mathematics. This position provided Newton the time to collect his notes and properly publish his work—a task he was always tardy on.

Shortly after his election to the Royal Society (in 1671), Newton published his first scientific paper. While an undergraduate, Newton had used a prism to refract a beam of white light into its primary colors (red, orange, yellow, green, blue, and violet). Newton reported this important discovery to the Royal Society. To Newton's surprise, this pioneering work was immediately attacked by Robert Hooke (1635–1703), an influential member of the society. Until his death, Hooke remained one of Newton's most fierce antagonists.

The attack was the first in a lifelong series of bitter disputes between Hooke and Newton. Generally, Newton only skirmished lightly then quietly retreated. This was Newton's lifelong pattern of avoiding direct conflict. When he became famous later in his life, Newton would start a controversy, withdraw, and then secretly manipulate others who would then carry the brunt of the battle against Newton's adversary. Newton's

ELASTIC BEHAVIOR OF A SPRING

When compressed or stretched within their elastic limits, springs store potential energy. In 1678, the British scientist Robert Hooke began studying the action of springs and reported that the extension (or compression) of an elastic material takes place in direct proportion to the force exerted on the material. Today, physicists use Hooke's law to quantify the displacement associated with the restoring force of an ideal spring. This restoring force *(F)* is expressed by the simple formula: $F = -kx$, where k is the spring constant and *x* is the displacement of the spring from its unconstrained length. Scientists use the minus sign to indicate that the restoring force always acts in a direction opposite to the displacement of the spring.

famous conflict with the German mathematician Gottfried Leibniz (1646–1716) over credit for the invention of calculus followed such a pattern. Calculus emerged as both Newton and Leibniz desired to understand motion and to accurately describe how motion-related variables like distance, velocity, and acceleration varied with respect to time. Through Newton's clever manipulation, the calculus controversy even took on nationalistic proportions as carefully coached pro-Newton British mathematicians bitterly argued against Leibniz and his supporting group of German mathematicians.

In August 1684, Halley traveled to Newton's home at Woolsthorpe. During this visit, Halley convinced the reclusive genius to address the following puzzle about planetary motion: What type of curve does a planet describe in its orbit around the Sun, assuming an inverse square law of attraction? To Halley's delight, Newton responded: "An ellipse." Halley pressed on and asked Newton how he knew the answer to this important question. Newton nonchalantly informed Halley that he had already done the calculations years ago (in about 1666). Since the absentminded Newton could not find his old calculations, he promised to send Halley another set as soon as he could.

To partially fulfill his promise, Newton later that year sent Halley his *De Motu Corporum* (On the motion of bodies). Newton's paper demonstrated that the force of gravity between two bodies is directly proportional to the product of their masses and inversely proportional to the square of the distance between them. (Physicists now call this physical relationship Newton's universal law of gravitation.) Halley was astounded and implored Newton to carefully document all of his work on gravitation and orbital mechanics. Through the patient encouragement and financial support of Halley, Newton published the *Principia* in 1687. The book contained his famous three laws of motion and the universal law of gravitation. The monumental work transformed physical science and completed the scientific revolution started by Copernicus, Kepler, and Galileo. Many scientists regard the *Principia* as the greatest technical accomplishment of the human mind.

Newton was extremely fragile. After completing the *Principia*, he drifted away from physics and astronomy and dabbled extensively in alchemy. In about 1693, he suffered a serious nervous disorder. Upon recovery, he left Cambridge (in 1696) and assumed a government post in London as warden (and later master) of the Royal Mint. During his years in London, Newton enjoyed power and worldly success. Robert Hooke,

NEWTON'S LAWS OF MOTION

Newton's laws of motion are the three fundamental postulates that form the basis of the mechanics of rigid bodies. He formulated these laws in about 1685, as he was studying the motion of the planets around the Sun. In 1687, Newton presented this work to the scientific community in the *Principia*.

Newton's first law of motion is concerned with the principle of inertia. It states that if a body in motion is not acted upon by an external force, the momentum remains constant. Scientists also refer to this law as the law of the conservation of momentum.

Newton's second law states that the rate of change of momentum of a body is proportional to the force acting upon the body and is in the direction of the applied force. A familiar statement of this law is the equation: $F = m\,a$, where F is the vector sum of the applied forces, m is the mass, and a represents the acceleration vector of the body.

Newton's third law is the principle of action and reaction. It states that for every force acting upon a body, there is a corresponding force of the same magnitude exerted by the body in the opposite direction.

his lifelong scientific antagonist, died in 1703. The following year (1704), the Royal Society elected Newton its president. Unrivaled, he won annual election to this position until his death. Newton was so bitter about his quarrels with Hooke that he waited until 1704 to publish his other major work, *Opticks*. Queen Anne knighted him in 1705.

Although his most innovative years were now clearly far behind him, Newton at this point in his life continued to exert great influence on the course of modern science. He used his position as president of the Royal Society to exercise autocratic (almost tyrannical) control over the careers of many younger scientists. Even late in life, he could not tolerate controversy. But now, as society president, he skillfully maneuvered younger scientists to fight his intellectual battles. In this manner, he continued to rule the scientific landscape until his death in London on March 20, 1727.

FORCE, ENERGY, AND WORK

The science of mechanics links the motion of a material object with the measurable quantities of mass, velocity, and acceleration. Through Newton's efforts, the term *force* entered the lexicon of physics. Scientists say a force influences a material object by causing a change in its state of motion. The concept of force emerges out of Newton's second law of motion. In its simplest and most familiar mathematical form, force *(F)* is the product of an object's mass (m) and its acceleration *(a)*—namely, $F = m\,a$. In his honor, scientists call the fundamental unit of force in the SI system, the newton (N). A force of one newton accelerates a mass of one kilogram at the rate of one meter per second per second. ($1\text{ N} = 1\text{ kg-m/s}^2$). In the American system, engineers define one pound-force (lbf) as the force equivalent to the weight of a one pound mass (1 lbm) object at sea level on Earth.

Physicists define energy (E) as the ability to do work. Within classical Newtonian physics, scientists describe mechanical work (W) as force *(F)* acting through a distance (d). The amount of work done is proportional to both the force involved and the distance over which the force is exerted in the direction of motion. A force perpendicular to the direction of motion performs no work. In the SI system, scientists measure energy with a unit called the joule (J). One joule represents a force of one newton (N) moving through a distance of one meter (m). This unit honors the British physicist James Prescott Joule (1818–89). (See chapter 4.) In the American system, engineers often express energy in terms of the British thermal unit (Btu). (One Btu equals 1,055 joules.)

In classical physics, energy (E), work (W), and distance (d) are scalar quantities; while velocity *(v)*, acceleration *(a)*, and force *(F)* are vector quantities. A scalar is a physical quantity that has magnitude only; while a vector is a physical quantity that has both magnitude and direction at each point in space.

The ability to quantify mechanical energy was a major step forward in understanding the behavior of the physical universe. Unraveling the mystery of heat was the next important milestone.

CHAPTER 4

Discovering the Nature of Heat

This chapter describes how the modern understanding of heat finally emerged in the 19th century after traveling down several misleading pathways. The two dominant incorrect interpretations were phlogiston theory and caloric theory. The phlogiston theory began in the 17th century and was later replaced in the 18th century by the caloric theory of heat. Both of these influential, but incorrect, models made an indelible imprint on the lexicon of the thermosciences. It took the persistence and hard work of several scientists in the 19th century, such as James Prescott Joule, to eventually displace caloric theory and replace it with the modern theory of heat, which relies upon both atomism and the notion of heat as a form of energy in transit.

EARLY THEORIES OF HEAT

This section describes how a study of combustion and gases helped scientists improve their understanding of heat. The phenomenon of heat has puzzled human beings since prehistoric times. Starting in the 17th century, alchemists and later the pneumatic chemists used phlogiston theory to explain combustion and the process of oxide formation. The term *pneumatic chemist* applies to the scientists of 17th through early 19th centuries, who studied chemical reactions and the physical behavior of gases in an effort to understand the true nature and composition of matter and energy.

Science historians identify the Flemish chemist and physician Johannes Baptista van Helmont (1579–1644) as the founder of pneumatic chemistry. Serving as a bridge between ancient alchemy and modern science, van Helmont stressed the importance of experiments in his chemical work, distinguished the existence of different gases in addition to atmospheric air, and coined the term *gas,* which he derived from the ancient Greek word for chaos.

In 1667, the German alchemist Johann Joachim Becher (1635–82) introduced a refined classification of substances (especially minerals). He replaced the air and fire portion of the four classical Greek elements with his own hypothesized three forms of earth. Becher called these new elemental substances: terra lapidea, terra fluida, and terra pinguis. For him, terra lapidea represented the principle of inertness in matter; terra fluida the principle of fluidity (that is, the volatile or mercurial behavior of matter); and terra pinguis the principle of combustibility. His concepts were actively promoted by the 18th-century German chemist Georg Ernst Stahl (1660–1734), who renamed terra pinguis and called it phlogiston—from

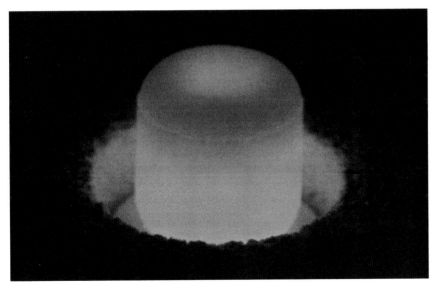

The radiant, thermal glow of an ingot of the radioactive isotope plutonium-238; with a half-life of 87.7 years, plutonium-238 serves as a reliable heat source on deep space missions within the solar system. Heat from the alpha-particle–emitting radioisotope is converted directly into electricity by means of radioisotope thermoelectric generators (RTGs). *(Department of Energy)*

the ancient Greek word φλογιστος, meaning fiery. Other 18th-century scientists also used phlogiston theory to explain combustion and the process of oxide formation.

According to this now-discarded theory, all flammable materials contained phlogiston, which was regarded as a weightless, colorless, and tasteless substance. During combustion, the flammable material released its phlogiston and what remained of the original material was a dephlogisticated substance—a crumby or calcined residual material called calx. Phlogistonists assumed that when the phlogiston left a substance during combustion it entered the air, which then became "phlogisticated air." Their pioneering combustion experiments with air in sealed containers revealed some very interesting, yet initially inexplicable, results.

Despite its shortcomings, phlogiston theory played a significant role in the emergence of chemistry as a science and in the overall understanding of gases, during its intellectual reign. For example, as a result of combustion experiments in 1772, the Scottish chemist Daniel Rutherford (1749–1819) isolated one type of "phlogisticated air" that he called "noxious air." At the time, Rutherford was a student of the famous Scottish chemist and phlogistonist Joseph Black (1729–99), who had previously discovered carbon dioxide (CO_2)—naming it "fixed air." Black was performing combustion experiments in which he would burn a candle in a sealed glass container and watch the flame eventually extinguish itself. He knew one of the products that remained was his "fixed air" (or carbon dioxide), but he was not sure what the other remaining gas was. So Black turned the problem over to Rutherford for further study.

Rutherford carefully extracted the carbon dioxide ("fixed air"), using various chemical absorbers and then tested the remaining gas. Since it would not support life, Rutherford called it "noxious air." At the time, he and other phlogistonists thought that during combustion, phlogiston left the original air (which supported combustion) and deposited itself in the gaseous combustion products. Since carbon dioxide did not support combustion, the remaining "noxious air" was assumed to contain the post-combustion phlogiston—thus the basic name, "phlogisticated air." Rutherford's leftover so-called noxious gas is now known as nitrogen—the abundant atmospheric gas that does not support respiration and does not experience combustion. Because of his experiments, Rutherford is generally credited with the discovery of nitrogen.

The French chemist Antoine Lavoisier (1743–94) performed carefully arranged mass-balance experiments in which he demonstrated

LAVOISIER AND THE RISE OF MODERN CHEMISTRY

The French scientist Antoine-Laurent Lavoisier founded modern chemistry. He was born into a wealthy Parisian family in 1743 and educated at the Collège Marzin. Although trained as a lawyer, his interest in science soon dominated his lifelong pursuits. A multitalented individual, from the 1760s onward he conducted scientific research, while also being involved in a variety of political, civic, and financial activities.

In 1768, Lavoisier became a member of Ferme Générale—a private, profit-making organization that collected taxes for the royal government. Although this financially lucrative position opened up other political opportunities for the brilliant scientist, the association with tax-collecting for the king would prove fatal. Lavoisier began an extensive series of combustion experiments in 1772, the primary purpose of which was to overthrow phlogiston theory. His careful analysis of combustion products allowed Lavoisier to demonstrate that carbon dioxide formed. He advocated the caloric theory of heat and suggested that combustion was a process in which the burning substance combined with a constituent of the air—a gas he called oxygine (a term meaning acid maker).

As part of his political activities, Lavoisier received an appointment in 1775 to serve as one of the commissioners at the arsenal in Paris. In this official capacity, he assumed responsibility for the production of gunpowder at the arsenal. Lavoisier's duties encouraged him to construct an excellent laboratory at the arsenal and to establish professional contacts with scientists throughout Europe.

One of Lavoisier's contributions to the establishment of modern chemistry was his creation of a system of chemical nomenclature, which he summarized in the 1787 publication *Method of Chemical Nomenclature.* Lavoisier's book promoted the principle by which chemists could assign every substance a specific name, based on the elements of which it was composed. His nomenclature system allowed scientists throughout Europe to quickly communicate and compare their research—thus enabling the rapid advance in chemical science during that period. Modern chemists still use Lavoisier's nomenclature.

In 1789, Lavoisier published his *Treatise of Elementary Chemistry.* Science historians consider it the first modern textbook in chemistry. In this book, Lavoisier discredited the phlogiston theory of heat and championed the caloric

(continues)

(continued)

theory. He advocated the important law of conservation of mass and suggested that an element was a substance that could not be further broken down by chemical means. Lavoisier also provided a list of all known elements in this book—including light and caloric (heat), which he erroneously regarded as elemental, fluidlike substances.

When the French Revolution took place in 1789, Lavoisier initially supported it, because he had previously campaigned for social reform. Throughout his life, he remained an honest, though politically naïve, individual. Lavoisier's privileged upbringing had socially insulated him from appreciating the growing restlessness of the peasants. Consequently, he did not properly heed the growing dangers around him and flee France. When the revolution became more extreme and transitioned into the Reign of Terror (1792–94), his membership in the much-hated royal tax-collecting organization (Ferme Générale) made him an obvious target for arrest and execution. All his brilliant contributions to the science of chemistry could not save his life. After a farcical trial, he lost his head to the guillotine in Paris on May 8, 1794.

that combustion involved the chemical combination of substances with oxygen. His work clearly dispelled the notion of phlogiston. In 1787, Lavoisier introduced the term *caloric* to support his own theory of heat—a theory in which heat was regarded as a weightless, colorless fluid that flowed from hot to cold. An incorrect concept, called the conservation of heat principle, served as one of the major premises in Lavoisier's caloric theory. Another important feature of this erroneous theory of heat was that the caloric fluid consisted of tiny, self-repelling particles that flowed easily into a substance, expanding the substance as its temperature increased.

In 1824, the French engineering physicist Nicolas Sadi Carnot (1746–1832) published *Reflections on the Motive Power of Fire* in which he correctly defined the maximum thermodynamic efficiency of a heat engine using caloric theory. Since steam engines dominated the Industrial Revolution, the thermodynamic importance of Carnot's work firmly entrenched the erroneous concept of heat as a fluid substance throughout the first half of the 19th century. It took pioneering experiments by the British-American scientist Benjamin Thompson (1753–1814) (later known as Count Rumford) and the British scientist and brewer James Prescott Joule (1818–89) to dislodge the caloric theory of heat.

Benjamin Thompson published an important paper in 1798 entitled *An Experimental Inquiry concerning the Source of Heat Excited by Friction.* He had conducted a series of experiments linking the heat released during cannon-boring operations in a Bavarian armory to mechanical work. He suggested that heat was the result of friction, rather than the flow of the hypothetical fluid substance called caloric. The more Thompson bore a particular cannon barrel, the more heat appeared due to mechanical friction—an experimental result that clearly violated (and thus disproved) the conservation of heat principle of caloric theory.

SADI CARNOT AND THE IDEAL HEAT ENGINE

Engineers define a heat engine as a thermodynamic system that receives energy in the form of heat and that, in the performance of an energy transformation on a working fluid, does work. Heat engines function in cycles. An ideal heat engine works in accordance with the Carnot cycle, while practical heat engines use thermodynamic cycles such as the Rankine, Brayton, Stirling, and Otto cycles. The steam engine, which stimulated the first Industrial Revolution, is a heat engine. The gasoline-fueled automobile engine is also a heat engine.

The Carnot cycle (shown in the illustration on page 66) is an idealized thermodynamic cycle for a theoretical heat engine. The basic concept of a heat engine was introduced by Nicolas Sadi Carnot in 1824. (Chapter 5 discusses the principles of thermodynamics.) The Carnot cycle consists of two adiabatic (no energy transfer as heat) stages, alternating with two isothermal (constant temperature) stages. Engineers use the pressure-volume (p-V) diagram to plot the various processes in the Carnot cycle. Two isothermal lines, shown with nominal values of $T_1 = 300$ and $T_4 = 200$ absolute temperature units (expressed as either rankines or kelvins) serve as the high (heat source) temperature and low (heat sink) temperature, respectively, for this ideal heat engine.

In stage one, the working fluid (arbitrarily chosen here as an ideal gas) experiences an isothermal expansion from thermodynamic state 1 to thermodynamic state 2. During the expansion process from state 1 to state 2, an amount of heat (called Q_1) is transferred from a source (at temperature T_1) to the expanding gas (assumed working fluid) in order to maintain a constant temperature. When the gas reaches the pressure (p_2) and volume

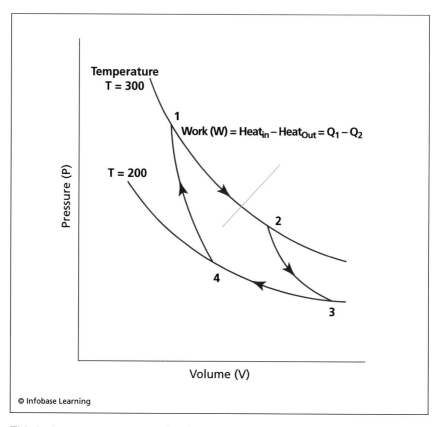

This is the pressure-volume (p-V) diagram for an ideal Carnot cycle. *(NASA)*

(V_2) conditions corresponding to thermodynamic state 2, its temperature (T_2) is still the same as the temperature as that found at state 1—namely, $T_1 = T_2 = 300$.

The second stage of the ideal Carnot cycle described here, involves an adiabatic (no heat transfer) expansion from thermodynamic state 2 to thermodynamic state 3. During this stage, the pressure of the gas (working fluid) decreases, the temperature decreases (to $T_3 = 200$), while the volume increases. As the working fluid travels along the path from state 2 to state 3, the expanding gas completes the power stroke of the Carnot cycle; the work of the expansion causes the temperature of the working fluid to fall from T_2 to T_3.

During the next two stages of the ideal Carnot cycle, the working fluid returns to its original thermodynamic state. In stage three, the gas experiences an isothermal compression, as it goes from state 3 to state

4. To keep the temperature constant during this compression process, an amount of heat (Q_2) is rejected to the surroundings (environment) at temperature ($T_3 = T_4 = 200$). During this isothermal compression process, the volume of the gas decreases from V_3 to V_4. The fourth and last stage of the ideal Carnot cycle involves adiabatic compression of the working fluid. The pressure of the gas increases and the volume of the gas decreases, as the temperature of the gas increases to T_1. At the end of the fourth stage, the gas is back at its original thermodynamic state (namely, p_1, V_1, and T_1). Functioning in a cycle, the working fluid continuously performs mechanical (expansion) work, as heat is added to and rejected from the engine.

During the Carnot cycle, an amount of work (W) has been produced by the gas, as it accepts an amount of heat (Q_1) from a source and rejects a lesser amount of heat (Q_2) to the surroundings. The fact that some heat has to be rejected to the environment is a consequence of the second law of thermodynamics. (See chapter 5.) Even though he used an erroneous concept of heat as a fluid (caloric theory), Sadi Carnot nevertheless correctly expressed the overall energy balance of an ideal heat engine as follows:

Mechanical Work$_{out}$ (W) = Heat Added (Q_1) – Heat Rejected (Q_2).

He also described the maximum thermal efficiency (symbol η_{th}) of an ideal heat engine using the basic equation: $\eta_{th} = 1 - T_{sink}/T_{source}$. In the ideal Carnot cycle described here, the heat sink temperature (T_{sink}) corresponds to $T_3 = T_4 = 200$; while the heat source temperature corresponds to $T_1 = T_2 = 300$. (Note that these temperatures must be expressed in absolute temperature units.) For the Carnot cycle examined here, the maximum thermal efficiency of an ideal heat engine operating between $T_{source} = 300$ and $T_{sink} = 200$ is 33.33 percent.

Mechanical engineers understand that the thermal efficiency of the Carnot cycle is the best possible efficiency of any heat engine operating between two temperature limits, T_{source} and T_{sink}. They sometimes refer to this fact as the Carnot principle. To achieve the maximum efficiency in a reversible heat engine, the heat source temperature should be as high as possible and the heat sink (environmental) temperature should be as low as possible. Heat engines operating on Earth must reject thermal energy to the surrounding terrestrial environment; so the sink temperature is limited to about 540 R (300 K)—depending upon local environmental conditions. Although the Carnot cycle is a postulated, idealized cycle, it is very useful in understanding the performance limits of real heat engines that operate using somewhat less than idealized Rankine, Brayton, and Otto cycles.

JAMES PRESCOTT JOULE

The British scientist and brewer James Prescott Joule was born into a comfortable merchant family and enjoyed private tutoring as a child. Joule's later interest in atomism can be directly attributed to the fact that the British schoolteacher and chemist John Dalton served as one of his tutors.

Like many other scientists in the first half of the 19th century, Joule became intrigued by the nature of energy and the transformation of energy from one form into another. In helping operate the family's brewery, he became familiar with some of the best available instruments to make accurate temperature measurements. At this time in Europe, the caloric theory of heat dominated the scientific landscape. Joule's pioneering experimental activities would play a major role in overthrowing caloric theory and guiding scientists to the modern interpretation of heat. Unfortunately, since he was a brewer by trade and not a formal member of the scientific establishment, much of his early, self-funded research was initially ignored.

Joule used his own private laboratory to quantitatively investigate the nature of heat. In 1840, he meticulously examined the rate at which electrical energy was converted into heat by an electric current as it passed through a wire. His experiments indicated that heat produced in a wire per second was proportional to the electric current squared (I^2) times the resistance of the wire (R). Engineers and scientists often use Joule's law of heating—namely, $P = I^2 R$, where P is heating rate (power) expressed in watts, I is current in amperes, and R is resistance in ohms.

Over the next several years Joule conducted a variety of interesting experiments to determine the mechanical equivalence of heat. Through meticulous measurements, he demonstrated the very important relationship between heat and mechanical work. His most successful efforts involved a system of weights (masses), a paddle wheel, and water placed in a carefully designed calorimeter. As the masses descended due to Earth's gravity, the paddle wheel turned, performing mechanical work on the water—thereby heating it. Joule precisely measured the subtle rise in water temperature. The modest temperature increase provided him experimental evidence of the equivalence of mechanical energy (work) and heat. In 1843, Joule reported that heat was another form of energy, like mechanical work. This simple experiment represents one of the greatest intellectual breakthroughs in classical physics. Joule published the results of his experiments in 1845 in a paper entitled "On the Existence of an Equivalent Relation between Heat and the Ordinary Forms of Mechanical Power."

Today, American engineers use the following values to express the relationship between heat and mechanical energy (work): one British thermal unit (Btu) equals 778.16 foot-pounds force (ft-lbf). Similarly, scientists

CONSERVATION OF ENERGY PRINCIPLE

One of the foundations of classical thermodynamics is the conservation of energy principle, which scientists often express as the first law of thermodynamics. While thinking about the relationship between biological processes and animal heat, the German physician and physicist Julius Robert von Mayer (1814–78) made an estimate of the mechanical equivalence of heat. In 1842, he published this work and his thoughts about the conservation of energy.

Independent of Joule, Mayer was the first person to suggest that energy can neither be created nor destroyed but only transformed. Although Mayer's calculations about the mechanical equivalence of heat preceded those of Joule, the latter generally receives credit for this important observation. The historic oversight is possibly due to the fact that Mayer did not present his ideas adequately to the scientific community.

An international dispute arose within the scientific community over assigning proper credit for measuring the mechanical equivalence of heat. Although historically justified, Mayer's claim on priority was generally ignored. In 1847, the German physician and physicist Hermann Ludwig Ferdinand von Helmholtz (1821–94) published the important work "On the Conservation of Force." In this document, Helmholtz drew upon the previous work of Carnot, the French engineer Émile Clapeyron (1799–1864), and Joule and then postulated that heat, magnetism, electricity, and mechanical work were all manifestations of energy. Since the scientific concept of energy was still emerging, Helmholtz stayed within the lexicon of the times by using the word force. His 1847 publication is often cited as the first scientific declaration of the conservation of energy principle and the earliest formal recognition of the first law of thermodynamics. Inexplicably, Helmholtz did not mention the earlier work of Mayer.

express the relationship between heat and work as follows: one calorie (cal) equals 4.1868 newton-meter (N-m).

Although initially met with resistance, Joule's discovery of the mechanical equivalence of heat was eventually recognized as a meticulous experimental demonstration of the conservation of energy principle—one of the intellectual pillars of thermodynamics. His experimental results represent one of the great breakthroughs in science. In honor of his accomplishments, the scientific community named the SI unit of energy the joule (J). Scientists define one joule (J) as the unit of energy equivalent to the work done when the point of application of a force of one newton moves a distance of one meter in the direction of the force.

Between 1853 and 1862, Joule collaborated with the Scottish physicist Lord Kelvin (William Thomson [1824–1907]). One product of their research was the discovery of the Joule-Thomson effect. This phenomenon involves the change (decrease) in temperature that occurs when a real gas expands through a throttling device, such as a porous plug or expansion valve, into a lower-pressure regime. Their combined efforts provided an experimental approach for scientists to determine other thermodynamic properties of interest.

SOME IMPORTANT CONCEPTS ASSOCIATED WITH HEAT

Heat transfer and thermodynamics play major roles in understanding the behavior of all materials. This section introduces some of the basic concepts used by scientists to explain the nature of heat and the thermodynamic behavior of materials. Unfortunately, some terms, such as *heat capacity* and *latent heat*, are potentially confusing. These terms appeared in the 18th century during the development of now-discarded theories of heat; yet they remain as part of today's scientific lexicon because of tradition.

From the late 19th century forward, scientists no longer viewed heat as an intrinsic property of matter, as has previously been thought by followers of the phlogiston theory, nor as a special, conserved fluid, as was incorrectly assumed in caloric theory. Rather, they began to recognize that heat, like work, is a form of energy in transit. On a microscopic scale, it is sometimes useful to envision heat as disorganized energy in transit—that is, the somewhat chaotic processes taking place as molecules or atoms randomly experience more energetic collisions and vibrations under the influence of temperature gradients.

Today, scientists recognize that there are three basic forms of heat transfer: conduction, convection, and radiation. Conduction is the transport of heat through an object by means of a temperature difference from a region of higher temperature to a region of lower temperature. The primary mechanism by which heat travels (conducts) through an object is the motion (vibration) of atoms and molecules. In liquids and gases, thermal conduction is due to molecular collisions. For solids, atomic vibrations are generally responsible for heat conduction. Higher temperature molecules and atoms vibrate in place more energetically, thereby transferring energy to adjacent molecules or atoms in lower temperature regions of the solid. For electrically conducting solids and liquid metals, thermal conduction is primarily accomplished by the migration of fast-moving electrons.

Convection is the form of heat transfer characterized by mass motions within a fluid, resulting in the transport and mixing of the properties of that fluid. One well-recognized example of convection is the up-and-down drafts in a fluid that is being heated from below while in a gravitational environment. Because the density of the heated fluid is lowered, the warmer fluid rises (natural convection); after cooling, the density of the fluid increases, and it tends to sink.

Finally, radiation (or radiant) heat transfer involves transfer of energy by the electromagnetic radiation that arises due to the temperature of a body. For objects found on Earth, most energy transfer of this type occurs in the infrared portion of the electromagnetic spectrum. (See chapter 7.) If the emitting object has a high enough temperature, it also will radiate in the visible portion of the spectrum and beyond. Scientists often use the term *thermal radiation* to distinguish this form of electromagnetic radiation from other forms, such as radio waves, light, X-rays, and gamma rays. Unlike convection and conduction, radiation heat transfer can take place in and through a vacuum.

There are two very important concepts associated with radiation heat transfer. First, every material substance in the universe has an absolute temperature (expressed in rankines or kelvins) associated with its energy content. Second, every material object radiates energy away at a rate proportional to the fourth power of its absolute temperature. Toward the end of the 19th century, the Austrian physicists Josef Stefan (1835–93) and Ludwig Boltzmann (1844–1906) collaborated on the formulation of an important physical principle, now called the Stefan-Boltzmann law. This physical law states that the energy radiated away per unit time by a blackbody is proportional to the fourth power of the object's absolute temperature. The blackbody represents the perfect emitter (and perfect absorber) of electromagnetic radiation. In nature, not every object behaves like a blackbody, but objects often approximate this behavior over certain temperature ranges and for certain radiating surface conditions. Radiation heat transfer is a very complex phenomenon, and further elaboration exceeds the scope of this section.

Thermal conductivity (symbol k) is a measure of a substance's ability to transport heat by means of conduction. Typical American customary units are Btu/(hr-ft-°F), and SI units are J/(s-m-°C). Metals, such as copper and silver, have high values of thermal conductivity; while foam materials have very low values of thermal conductivity. The thermal conductivity of silver at 80°F (27°C) is 248 Btu/(hr-ft-°F) (429 J/[s-m-°C]). In contrast, the thermal conductivity urethane rigid foam (insulation) at 80°F (27°C) is just 0.015 Btu/(hr-ft-°F) (0.026 J/[s-m-°C]). Clearly, some solid materials (called thermal conductors) transport energy as heat very well; other materials (called thermal insulators) resist the conduction of heat.

In the 18th century, the Scottish chemist Joseph Black introduced the concepts of specific heat and latent heat. Even though Black endorsed the erroneous phlogiston theory of heat, his insights into how different substances experienced different increases in temperature when they absorbed the same quantities of heat were quite accurate. Black also recognized that a solid substance required a special amount of heat addition (which he called the latent or hidden heat), while experiencing a change in state, when heated and completely melted at constant pressure under isothermal (constant temperature) conditions.

Scientists use specific heat (symbol c) in heat transfer analysis and thermodynamics as a measure of the heat capacity of a substance per unit mass. They define specific heat as the quantity of heat necessary to raise the temperature of a unit mass of substance by one degree (°F [°C]). At a temperature of 80°F (27°C) and a pressure of one atmosphere, silver (Ag) has a specific heat of 0.057 Btu/(lbm-°F) in American customary units or 0.234 kJ/(kg-°C) in SI units.

One practical use of the specific heat of a substance is to determine how much energy must be added as heat to a given mass of that substance to change its temperature from some reference state (called T_0) to a higher temperature state (called T_1). The process is called sensible heat addition, when the substance being heated does not undergo a change in state. Consider a 2.2 lbm (1 kg) mass of silver (m_{Ag}) initially at room temperature and one atmosphere pressure. Using the substance's specific heat, an engineer can calculate the amount of heat required (symbolized as ΔQ_{needed}) to raise the metal's temperature from $T_0 = 80°F$ (27°C) to $T_1 = 117°F$ (47°C).

The engineer uses the following simple equation: $\Delta Q_{needed} = c_{Ag} m_{Ag} (T_1 - T_0)$, where c_{Ag} is the specific heat of silver. After inserting appropriate values, the engineer determines that it will take about 4.55 Btu (4,800 joules) to raise the block of silver's temperature by 37°F (20°C).

Different materials have different values of specific heat. The specific heat of common red brick is about 0.20 Btu/(lbm-°F) or 0.84 kJ/(kg-°C) at a temperature of 80°F (27°C) and a pressure of one atmosphere. Using the same equation, the engineer calculates that it would take approximately 16 Btu (16,800 J) to raise the temperature of a 2.2 lbm (1 kg) mass brick by 37°F (20°C).

Not every time a solid is heated does its temperature rise. If the material is at the melting point, the addition of heat will cause the material to experience a change of state. Scientists define the melting point as the temperature (at a specified pressure) at which a substance experiences a change from the solid state to the liquid state. At this temperature the solid and liquid states of a substance can exist together in equilibrium. At one atmosphere pressure, the approximate melting points (expressed in absolute temperature units) of some solid substances are: sulfur 705.6 R (392 K); tin 909 R (505 K); lead 1,082 R (601 K); aluminum (pure) 1,679 R (933 K); silver 2,223 R (1,235 K); gold 2,405 R (1,336 K); copper (pure) 2,444 R (1,358 K); iron (pure) 3,258 R (1,810 K); titanium 3,515 R (1,953 K); and tungsten 6,624 R (3,680 K).

While investigating the transformation of solids into liquids (by melting) and liquids into gases (by boiling), Joseph Black coined the term *latent heat* to describe the heat energy flow in (or out) of a substance necessary to achieve the physical change of state. This term still lingers in thermodynamics, but has largely been replaced by the general term *heat of transformation* (symbol L), or the more rigorous term *enthalpy of transformation*. (Enthalpy [symbol H or h] is a somewhat complicated thermodynamic property of the substance or system under study.) The heat of transformation is the amount of heat per unit mass that must be added to (or removed from) a substance in order to produce a constant temperature (isothermal) change in physical state. Scientists define melting as the process whereby a solid substance transforms into a liquid. The heat of melting is the amount of heat that must be added to a unit mass of solid in order to change it completely into a liquid.

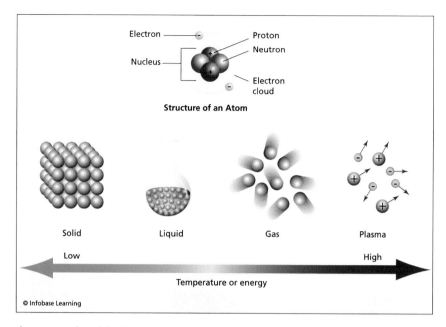

Structure of an Atom

As energy is added to a solid, its temperature increases and the solid becomes a liquid. Further addition of energy allows the liquid to become a gas. If the gas is heated a great deal more, its atoms break apart into charged particles, resulting in the fourth state of matter called plasma. *(based on NASA-sponsored artwork)*

Scientists call the reverse process freezing. The heat of fusion is the amount of heat that must be removed from a liquid while it changes (freezes) into a solid. Melting is an exothermic process, meaning heat must be added to the solid in order to give its molecules or atoms more energy to escape their rigid structure and move about more easily. Freezing is an endothermic process, meaning energy must be removed from a substance in the liquid state in order to slow down its atoms and molecules and allow them to form a rigid structure. With a melting point of 491 R (273 K), more commonly expressed as 32°F (0°C), ice has a heat of melting of 143 Btu/lbm (333 kJ/kg) at one atmosphere pressure. The heat of melting of lead is 9.98 Btu/lbm (23.2 kJ/kg); silver 45.2 Btu/lbm (105 kJ/kg); and copper 89 Btu/lbm (207 kJ/kg).

During isothermal (constant temperature) changes in state, the heat of fusion and the heat of melting have the same numerical values. What is different is the direction of heat flow into (for melting) or out of (for freezing) the substance. The term *melting point* is synonymous with the term *freezing point.* A liquid freezes when the necessary amount of heat is removed at the freezing point and the substance transforms into the solid state.

Certain substances undergo a transition directly from the solid state into the gaseous (vapor) state without passing through the liquid state. Scientists call this process sublimation. At one atmosphere pressure, solid carbon dioxide (commonly called dry ice) sublimes directly into gaseous carbon dioxide at a temperature of −109.6°F (350.4 R) or −78.5°C (194.65 K). The corresponding heat of sublimation is 245.5 Btu/lbm (571 kJ/kg). Scientists define deposition as the reverse process in which a gas directly transforms into a solid, without passing through the liquid state. Sublimation requires that heat be added to the solidified gas to transform it directly into a vapor; deposition required that the gas have heat removed in order to transform it directly into a solid. Under the proper atmospheric conditions, water vapor can directly transition into the solid state and appear as frost or snowflakes.

CHAPTER 5

Thermodynamics

A scientific understanding of the energy of matter involves thermodynamics, the branch of physics that deals with the flow of work and heat in response to various energy transformations. Classical thermodynamics emerged in the 19th century as scientists and engineers attempted to analyze and improve steam engines. Thermodynamics involves a few basic laws that describe the transformation of heat into other forms of energy, and vice versa. The chapter presents some of the fundamental principles and concepts associated with this important science. Classical thermodynamics uses just a few macroscopically measurable physical properties, such as mass, temperature, pressure, and volume to describe and predict the overall behavior of complex systems, like steam engines and gasoline engines. The chapter includes a discussion of four important thermodynamic cycles: the Rankine, Brayton, Stirling, and Otto cycles. The chapter also contains a brief discussion of the thermodynamics of chemical reactions and the explosive behavior of certain chemical compounds.

STEAM AND THE FIRST INDUSTRIAL REVOLUTION

Historians define the First Industrial Revolution as the period of enormous cultural, technical, and socioeconomic transformation that occurred in the late 18th and early 19th centuries. It served as a major stimulus for improvements in materials science and the production of advanced machines. Starting in Great Britain and quickly spreading throughout western Europe and North America, factory-based production and machine-dominated manufacturing began displacing economies based on manual labor. Historians often treat this transformation as being comparable to the Neolithic invention of farming with respect to the consequences on the trajectory of human civilization. Because of the global extent of the 19th-century British Empire, this wave of social and technical change quickly spread throughout the world.

British engineers and business entrepreneurs led the charge in the late 18th century by developing steam engines (fueled primarily by coal) and

Engineers use thermodynamics to design automotive engines. This picture shows a monster truck, named El Toro Loco, competing during the 2009 Monster Jam at the Alamodome in San Antonio, Texas. Monster trucks have big block supercharged racing V-8 engines that churn out more than 1,400 horsepower (1,044,400 watts). Each vehicle has a mass of about 10,000 lb (4,545 kg) and is equipped with four-wheel steering. The tires are 66 inches (168 cm) in diameter. *(U.S. Air Force)*

using steam-powered machinery in manufacturing (initially in the textile industry). These technical innovations were quickly followed by the development of all metal machine tools in the early 19th century. The availability of metal machine tools led to the development of more advanced machines for use in factory-based manufacturing in other industries. More and less expensive goods became available. Workers left the farms and flocked to urban areas to work in factories.

THE POWER OF STEAM

As part of the first Industrial Revolution, Great Britain led the world in the development of railroads and steam-powered locomotives in the early 19th century. One dominant individual was the British inventor George Stephenson (1781–48). The Stephenson track gauge of four feet and 8.5 inches (1.453 m) became the standard gauge for many of the world's railroads. Railroad engineers define a rail gauge as the distance between the inside edges of the rails.

Across the Atlantic Ocean, railroads greatly assisted the rapid development of the United States. For example, the citizens of Chicago witnessed their first steam-powered locomotive in 1848, obtained rail service to the eastern portion of the United States by 1852, and eight years later enjoyed service by 11 railroads with more than 100 train arrivals daily. When Abraham Lincoln (1809–65) returned to Springfield, Illinois, in March 1849 after serving his single term in Congress (1847–49), the trip via stagecoach, steamboat, and rail took him 11 days. By 1861, an all-rail trip over several alternative routes from Springfield to Washington would take less than two days.

One of the major events in American transportation history occurred at Promontory Summit, Utah, on May 10, 1869. The date marks the nominal completion of the country's first transcontinental railroad. Using a golden spike, officials ceremoniously completed the final section of track, linking the Central Pacific and Union Pacific railroads. As part of the pageantry, railroad officials located the two participating locomotives, Union Pacific Locomotive No. 119 and Central Pacific Locomotive No. 60 (also called *Jupiter*), face-to-face—separated by only the width of a single railroad tie.

Civil War veterans and Irish immigrants served as the primary workers on the Union Pacific rail route, which extended about 1,087 miles (1,750 km) from Council Bluff, Iowa, to Promontory Summit, Utah (then a territory). Mostly Chi-

A common misperception is that the Scottish engineer James Watt invented the steam engine. Actually, Watt was working as an instrument maker at the University of Glasgow in 1764, when officials at that institution gave him a Newcomen engine to repair. The repair project introduced Watt to the world of steam, and the inventive young man soon not only repaired the Newcomen engine but greatly improved its design. Watt's stroke of genius was a decision to add a separate condenser to the

nese immigrants labored on the route completed by the Central Pacific railroad. They worked hard laying approximately 690 miles (1,111 km) of track from Sacramento, California, across the rugged Sierra Nevada, and on to Promontory Summit.

In November 1869, the Central Pacific completed tracks that connected Sacramento to the San Francisco Bay Area at Oakland. By 1870, steam-powered locomotives allowed passengers to comfortably travel from the Missouri River (at Council Bluffs, Iowa) to San Francisco Bay in about six days. Previously, overland pioneers had to spend four to six danger-filled months to make the same journey traveling mostly in horse- or oxen-drawn covered wagons. After the completion of the transcontinental railroad, the power of steam began forging the East and West Coast regions into a single nation.

A replica of Union Pacific Locomotive No. 119 puffs steam during a winter celebration at the Golden Spike National Historic Site, Promontory Summit, Utah. *(National Park Service)*

Today, the Golden Spike National Historic Site of the National Park Service commemorates this famous event. Operational replicas of Central Pacific Locomotive No. 60 *(Jupiter)* and the Union Pacific Locomotive No. 119 participate in various celebrations. The site is approximately 32 miles (51 km) from Brigham City, Utah. Themselves victims of progress, the original engines were scrapped at the beginning of the 20th century.

heat engine, thereby greatly improving its thermodynamic efficiency. The British engineer Thomas Newcomen (1663–1729) developed and patented the original Newcomen engine at the start of the 18th century. The primary application of this bulky and thermally inefficient heat engine was to operate as a steam-powered vacuum engine that could drain water out of deep mines.

In 1775, Watt entered a business partnership with the British entrepreneur Matthew Boulton (1728–1809). Continually being improved, the Watt steam engine soon dominated mechanical power production all over the United Kingdom and eventually throughout much of continental Europe and North America. Steam provided power for factories, which no longer had to be located near sources of falling water (hydropower). Other versions of the Watt steam engine began to power ships and propel railroad trains. In a few short years, steam-powered engines transformed human civilization.

As scientists and engineers sought to make steam engines more efficient, they began to explore the nature of heat and its relationship to mechanical work. They also investigated the thermal properties of matter, giving rise to the science of thermodynamics. Steam not only powered the first Industrial Revolution, it also powered an amazing intellectual revolution in the 19th century, involving materials science, fluid mechanics, thermodynamics, and the rebirth of atomic theory.

BASIC MACROSCOPIC PROPERTIES USED IN THERMODYNAMICS

This section introduces the three important macroscopic properties of matter (density, pressure, and temperature) that are the building blocks of thermodynamic analyses. By thinking about the atoms that make up different materials, scientists can quantify and predict how interplay at the atomic (microscopic) level results in the physical properties that are measurable on a macroscopic scale.

Density

To assist in more easily identifying and characterizing different materials, scientists devised the material property called density. Solid matter is generally denser than liquid matter; liquid matter denser than gases. Scientists define density as the amount of mass contained in a given volume. They frequently use the lower case Greek letter rho (ρ), as the symbol for density in technical publications and equations.

Scientists use the density (ρ) of a material to determine how massive a given volume of that particular material is. Furthermore, equal volumes of different materials usually have different density values. Density is a function of both the atoms from which a material is composed and how closely packed the atoms are arranged in the particular material. At room temperature (nominally 68°F [20°C]) and one atmosphere pressure, the density of some familiar materials is as follows: gold 1,205 lbm/ft³ (19,300 kg/m³ [19.3 g/cm³]); iron 493 lbm/ft³ (7,900 kg/m³ [7.9 g/cm³]); diamond (carbon) 219 lbm/ft³ (3,500 kg/m³ [3.5 g/cm³]); aluminum 169 lbm/ft³ (2,700 kg/m³ [2.7 g/cm³]); and water 62.3 lbm/ft³ (997.4 kg/m³) [0.997 g/cm³]). Like most gases at room temperature and one atmosphere pressure, oxygen has a density of just 0.083 lbm/ft³ (1.33 kg/m³ [1.33 × 10⁻³ g/cm³])—a value that is about 1,000 times lower than the density of most solid or liquid materials normally encountered on Earth's surface.

Most materials have a higher density when solid, than when liquid. However, the maximum density of water (pure) occurs when the liquid's temperature is at 39.16°F (3.98°C). Upon freezing, water transforms to ice, which has a lower value of density—namely, 56.81 lbm/ft³ (910 kg/m³ [0.91 g/cm³]) at 32°F (0°C). This decrease in density upon freezing is the reason why ice floats on water—a very important, yet unusual, natural phenomenon.

Unlike rigid solids, fluids are materials that can flow. So, engineers use pressure differentials to move fluids. They design pumps to move liquids (often treated as incompressible); while they design fans to move compressible fluids (gases). An incompressible fluid is assumed to have a constant value of density; a compressible fluid has a variable density. One of the interesting characteristics of gases is that, unlike solids or liquids, they can be compressed into smaller and smaller volumes.

Scientists use either the density or specific volume of a gas to describe its thermodynamic state. Engineers prefer to use the density of the gas as the intensive property associated with mass, because in many gas dynamic problems the mass of the gas varies from location to location.

Scientists know that the physical properties of matter are often interrelated—namely, when one physical property (like temperature) changes, other physical properties (like pressure or density) also change. As a direct result of the Scientific Revolution, people learned how to define the behavior of materials by developing special mathematical expressions, termed *equations of state*. Scientists created these mathematical relationships, using both theory and empirical data from many carefully conducted laboratory experiments.

SPECIFIC VOLUME

In many thermodynamic problems, engineers find it necessary to include the density of the working fluid. To assist in their computations, they define an intensive macroscopic property, called the specific volume (usual symbol v). Specific volume is the volume per unit mass of a substance. It is the reciprocal of density and has the following units: ft^3/lbm in American units and m^3/kg in SI units.

An intensive thermodynamic property is independent of the mass involved in a particular problem; an extensive thermodynamic property varies directly with the mass and is dependent upon the actual size (or extent) of the system under study. The total volume (V) of a system is an example of an extensive property, while the specific volume *(v)* is an intensive property.

The use of intensive properties facilitates the development of thermodynamic property tables and the comparison of various physical processes with just a few well-understood thermodynamic laws and principles. For example, the density of dry air at 68°F (20°C) and one atmosphere pressure is 0.0755 lbm/ft^3 (1.21 kg/m^3); therefore, the specific volume of dry air under the same conditions of temperature and pressure is 13.238 ft^3/lbm (0.826 m^3/kg). If engineers raise the pressure of this sample of air to 50 atmospheres, while keeping the temperature at 68°F (20°C), the density becomes 3.777 lbm/ft^3 (60.5 kg/m^3) and the specific volume is 0.265 ft^3/lbm (0.0165 m^3/kg).

Pressure

Scientists describe pressure (P) as force per unit area. The most commonly encountered American unit of pressure is pounds-force per square inch (psi). In SI units, the fundamental unit of pressure is termed the pascal (Pa) in honor of Blaise Pascal (1623–62). This 17th-century French scientist conducted many pioneering experiments in fluid mechanics. One pascal represents a force of one newton (N) exerted over an area on one square meter—that is, $1 Pa = 1 N/m^2$. One psi is approximately equal to 6,895 Pa.

Temperature

While temperature is one of the most familiar physical variables, it is also one of the most difficult to quantify. Scientists suggest that on the macroscopic scale temperature is the physical quantity that indicates how hot or cold an object is relative to an agreed-upon standard value. Temperature defines the natural direction in which energy will flow as heat—namely from a higher temperature (hot) region to a lower temperature (cold) region. Taking a microscopic perspective, temperature indicates the speed at which the atoms and molecules of a substance are moving.

Scientists recognize that every object has the physical property called temperature. They further understand that when two bodies are in thermal equilibrium, their temperatures are equal. A thermometer is an instrument that measures temperatures relative to some reference value. As part of the Scientific Revolution, creative individuals began using a variety of physical principles, natural references, and scales in their attempts to quantity the property of temperature.

In about 1592, Galileo attempted to measure temperature with a device he called the thermoscope. Although Galileo's work represented the first serious attempt to harness the notion of temperature as a useful scientific property, his thermoscope (while innovative) did not supply scientifically significant temperature data.

Thermometry is an important part of thermodynamics and heat transfer studies. This figure provides a close-up view of a portion of an analog thermometer calibrated to simultaneously display relative temperature values in both Fahrenheit (left) and Celsius (right) degrees. *(NIST)*

The German physicist Daniel Gabriel Fahrenheit (1686–1736) was the first person to develop a thermometer capable of making accurate, reproducible measurements of temperature. In 1709, he observed that alcohol expanded when heated. He constructed the first closed-bulb glass thermometer with alcohol as the temperature-sensitive working fluid. Five years later (in 1714), he used mercury as the thermometer's working fluid. Fahrenheit selected an interesting three-point temperature reference scale for his original thermometers. His zero point (0°F) was the lowest temperature he could achieve with a chilling mixture of ice, water, and ammonium chloride (NH_4Cl). Fahrenheit then used a mixture of just water and ice as his second reference temperature (32°F). Finally, he chose his own body temperature (recorded as 96°F) as the scale's third reference temperature.

After his death, other scientists revised and refined the original Fahrenheit temperature scale, making sure there were 180 degrees between the freezing point of water (32°F) and the boiling point of water (212°F) at one atmosphere pressure. On this refined scale, the average temperature of the human body appeared as 98.6°F. Although the Fahrenheit temperature scale is still used in the United States, most of the other nations in the world have adopted another relative temperature scale, called the Celsius scale.

In 1742, the Swedish astronomer Anders Celsius (1701–44) introduced the relative temperature scale that now carries his name. He initially selected the upper (100 degree) reference temperature on his new scale as the freezing point of water and the lower (0 degree) reference temperature as the boiling of water at one atmosphere pressure. He then divided the scale into 100 units. After Celsius's death, the Swedish botanist and zoologist Carl Linnaeus (1707–78) introduced the present-day Celsius scale thermometer by reversing the reference temperatures. The modern Celsius temperature scale is a relative temperature scale in which the range between two reference points (the freezing point of water at 0°C and the boiling point of water at 100°C) are conveniently divided into 100 equal units or degrees.

Scientists describe a relative temperature scale as one that measures how far above or below a certain temperature measurement is with respect to a specific reference point. The individual degrees or units in relative scale are determined by dividing the relative scale between two known reference temperature points (such as the freezing and boiling points of water at one atmosphere pressure) into a convenient number of temperature units (such as 100 or 180).

Despite considerable progress in thermometry in the 18th century, scientists still needed a more comprehensive temperature scale—namely one that included the concept of absolute zero, the lowest possible temperature at which molecular motion ceases. The Irish-born Scottish physicist William Thomson (Lord Kelvin) proposed an absolute temperature scale in 1848. The scientific community quickly embraced Kelvin's temperature scale. Scientists generally use absolute temperatures in such disciplines as physics, astronomy, and chemistry; engineers use either relative or absolute temperatures in thermodynamics, heat transfer analyses, and mechanics, depending upon the nature of the problem. Absolute temperature values are always positive; but relative temperatures can have positive or negative values.

RELATIONSHIP BETWEEN FLUID FLOW AND THERMODYNAMICS

Engineers define a working fluid as the liquid or gas they employ as the medium for the transfer of energy from one part of a thermodynamic system to another. As discussed in subsequent sections of this chapter, the working fluid plays a very important role in heat engine cycles, such as the Rankine cycle and the Brayton cycle.

This section presents several basic concepts used in the thermodynamic analysis of flowing fluids. If scientists can assume that the density of the fluid is constant, they can treat the fluid as incompressible. If the density of a fluid experiences significant changes, scientists must treat the fluid as compressible. Scientists understand that analyzing the high velocity flow of a compressible fluid (like the hot combustion gases expanding through a jet engine's exhaust nozzle) is a much more complicated task than analyzing the flow of an incompressible fluid (such as water) through a pipe.

A fundamental idea in compressible fluid flow is that of the continuum. As part of the Scientific Revolution, scientists began treating a flowing fluid as a continuous material, or continuum. Scientists now recognize that the approach is valid so long as the smallest volume of interest in a problem contains a sufficient number of atoms or molecules to make statistical averaging meaningful. The major advantage of the continuum approximation is that scientists can describe the complex behavior of innumerable individual atoms or molecules with a few macroscopic properties that quantify the gross physical behavior of the substance. When scientists deal with compressible fluids and treat them as continua, they typically use the following macroscopic thermodynamic properties: density, pressure, temperature, viscosity, velocity, internal energy, enthalpy, entropy, and thermal conductivity.

As part of the continuum approximation, scientists and engineers assume that at any instant, every point of a fluid continuum has a corresponding fluid velocity vector. Instantaneous streamlines of flow are the curves that are tangent everywhere to the velocity vector. Streamlines are very useful in helping scientists visualize flow patterns. In steady flow, the streamlines are constant and represent the trajectories (path lines) taken by fluid particles. In unsteady flow, the streamlines change continuously.

The following thermodynamic properties are now briefly discussed: internal energy, enthalpy, and entropy. Internal energy (symbol: u) is an intrinsic, macroscopic thermodynamic property interpretable through statistical mechanics as a measure of the microscopic energy modes (that is, molecular activity) of a system. This important property is associated with the first law of thermodynamics.

Enthalpy (symbol: h) is an intrinsic property of a thermodynamic system best described by its defining formula: $h \equiv u + P\,v$, where u is internal energy, P is pressure, and v is specific volume. Unfortunately, while very useful in the study of open systems and fluid flow problems, this thermodynamic property does not lend itself to a simple physical interpretation. However, as an intrinsic property of a thermodynamic system, specific enthalpy has the dimensions of energy per unit mass (Btu/lbm [J/kg]) and provides a macroscopic measure of the internal energy content and flow work performed on/by a fluid as it travels through a thermodynamic system.

Many technical papers and books have been written about the thermodynamic property, called entropy (symbol: s). Entropy is an intrinsic thermodynamic property that serves as a measure of the extent to which the energy of a system is unavailable. Based on the second law of thermodynamics, specific entropy has the dimensions of energy per unit mass per degree of absolute temperature; typical units are Btu/(lbm-R) or J/(kg-K). In the late 19th century, as part of the development of kinetic theory, the Austrian physicist Ludwig Boltzmann and the American scientist Josiah Willard Gibbs (1839–1903) introduced a statistical definition of entropy as a disorder or uncertainty indicator.

As mentioned in chapter 4, thermal conductivity (symbol: k) is an intrinsic physical property of a substance that describes its ability to conduct heat (that is, transport energy) as a consequence of molecular motions in gases and liquids and as a combination of lattice vibrations and electron transport in solid substances. In 1822, the French mathematician Jean-Baptiste-Joseph Fourier (1768–1830) examined steady state heat transfer through a solid rod and provided a mathematical description of conduction heat transfer. He equated energy flow as heat as being proportional to the temperature gradient in the direction of heat flow and the area perpendicular to the heat flow. The constant of proportionality in his famous partial differential equation became known as the thermal conductivity (k) of the substance. Using mathematics, scientists express the one-dimensional, steady state version of this law as: $q'' = -k\,\partial T/\partial x$, where q'' is the heat flux and $\partial T/\partial x$ is the temperature gradient. Fourier's law of heat conduction provided the first significant macroscopic measure of the rather elusive microscopic phenomenon of heat transfer by atomic motions.

In thermodynamics and gas dynamics, scientists make considerable use of the closed and open systems. A closed system is a thermodynamic system in which no transfer of mass takes place across the system's boundaries. However, energy in the form of heat and work may cross the system's boundaries. A frequently used closed mass system model involves the piston cylinder. The basic thermodynamic system is the collection of working fluid (gas) contained in the piston-cylinder arrangement. As the piston pushes in (that is, does work on the gas), the gas contained in the cylinder experiences a decrease in the volume and an increase in pressure. (Scientists assume the piston is properly sealed so the gas cannot escape while being compressed.) If the piston is both rigid and well-insulated, heat will not be able to leave the compressed gas and its temperature will rise. Scientists use the term *adiabatic* to describe a process in which there is no energy transfer as heat across the boundaries of a thermodynamic system. They often treat the gas in the cylinder as an ideal gas, which in this example is experiencing an adiabatic compression. If the cylinder is cooled during the compression process in such a way as to keep the compressed gas at its initial temperature, then scientists say the gas is undergoing an isothermal compression.

The gas in the piston-cylinder example provides an excellent opportunity to discuss the very important state principle of thermodynamics. The state principle assumes that the number of independent properties needed to define the thermodynamic state of a system is equal to the number of possible work modes plus one. Scientists define a simple substance as one that has only one possible work mode. The gas in the piston-cylinder assembly has one work mode, namely volume change (ΔV). This means that scientists need only two independent properties to describe the state of the gas, while it is being compressed or expanded. They often select two of the following properties: temperature, pressure, or specific volume.

The concept of a pure substance is also important in thermodynamics. Scientists define a pure substance as one that is homogeneous and maintains the same chemical composition in all its common physical states (that is, as a solid, liquid, or gas). Water is an example of a pure substance, since it maintains the same chemical composition whether it is ice, liquid water, or water vapor (steam). Scientists often treat mixtures of gases, like air, as a pure substance. If a gaseous mixture is cooled too much, it can reach a sufficiently low temperature so that some of its component gases condense and become liquids. At that point, the original gaseous mixture can no longer be treated as a pure substance, because its chemical composition has changed. Unless noted otherwise, this book treats both air and water as simple substances and as pure substances.

The Swiss mathematician Leonard Euler (1707–83) and the Italian-French mathematician Joseph-Louis Lagrange (1736–1813) provided scientists the tools necessary to describe the motion of fluids in terms of a continuum. The method of Euler focuses upon a fixed point in space and specifies the density, pressure, temperature, etc., of the fluid particle that happens to occupy that point at each instant of time. Scientists prefer to use the Eulerian method when they solve problems involving fluid motion. The method of Lagrange involves the history of individual particles as they move through a system. In the Lagrangian method, the instantaneous density, pressure, temperature, state of stress, etc., of a fluid particle of fixed identity are specified at each instant of time. Scientists use the Lagrangian description in solid mechanics.

IDEAL GAS EQUATION OF STATE

The ideal (perfect) gas equation of state is an important principle in thermodynamics. This principle states that the pressure (P), absolute temperature (T), and total volume (V) of a gas are related as follows: $P V = N R_U T$, where N is the number of moles of gas and R_U is the universal gas constant. For any gas, the universal gas constant (R_U) has the following value: 1545.3 ft-lbf/(lbmol-R) in American units and 8314.5 J/(kgmol-K) in SI units.

Engineers often prefer to express the ideal gas equation as: $P v = R T$, where v is the specific volume and R is the specific gas constant. At low pressures and moderate temperatures, many real gases approximate ideal gas behavior quite well. Assuming ideal gas behavior, air has a specific gas constant value (R) equal to 53.34 ft-lbf/(lbm-R), or 0.287 kJ/(kg-K), at one atmosphere pressure and 77°F(25°C).

When they apply the conservation of mass principle in classical fluid motion problems, scientists generally neglect relativistic effects and nuclear reactions. Invocation of this principle implies that the mass of the system is constant. Finally, scientists use Newton's second law of motion as the fundamental physical principle in the fluid dynamics. The momentum theorem states that the net force acting instantaneously on the gas (or liquid) within a control volume equals the time rate of change of momentum within that control volume plus the excess of outgoing momentum flux minus the incoming momentum flux.

The assumption of a gaseous fluid behaving as a continuum fails, whenever the mean free path of the molecules becomes comparable in size with the smallest significant physical dimension of the problem. For example, the continuum approach of classical gas dynamics is not appropriate for treating the highly rarified gases emitted by a rocket's nozzle, as the rocket operates at very high altitude above Earth. For such problems, engineers must pursue a solution in terms of the microscopic description of matter found in the kinetic theory.

FUNDAMENTAL LAWS OF THERMODYNAMICS

Thermodynamics is an elegant branch of physics that is based upon several fundamental laws. Scientists and engineers use thermodynamics to describe how heat and mechanical energy (work) interact with substances in various systems. A thermodynamic system is simply a collection of matter and space (specified volume) with its boundaries defined in such a way that energy transfer (as work and heat) across the boundaries can be identified and understood easily. The surroundings represent everything else that is not included in the thermodynamic system. Scientists use the term *steady state* to refer to a condition where the properties at any given point within a thermodynamic system remain constant over time. Neither mass nor energy accumulate (or deplete) in a steady state system.

A closed system is a system for which only energy (but not matter) can cross the boundaries. An open system can experience both matter and energy transfer across its boundaries. A control volume is a fixed region in space that is defined and studied as a thermodynamic system. Engineers often use the control volume in their analysis of open systems, such as jet engines and gas turbine power plants.

The zeroth law of thermodynamics states that two systems, each in thermal equilibrium with a third system, are in thermal equilibrium with each other. This statement is actually an implicit part of the concept of temperature. The law is important in the field of thermometry and in the establishment of empirical temperature scales.

The first law is the conservation of energy principle. For a control mass (that is, a thermodynamic system of specified matter), engineers express the first law in the form of the following basic energy balance: energy output − energy input = change in energy storage.

The second law is an inequality asserting that it is impossible to transfer heat from a colder to a warmer system without the occurrence of other simultaneous changes in the two systems or the surroundings (environment). Another way of stating the second law is that the total change of entropy (ΔS) for an isolated system is greater than or equal to zero. Scientists express this particular statement mathematically as $(\Delta S)_{isolated} \geq 0$. An isolated system can experience neither matter nor energy transfer across its boundaries.

The third law states that the entropy of any pure substance in thermodynamic equilibrium approaches zero as the absolute temperature approaches zero. The law is important in that it furnishes a basis for calculating the absolute entropies of substances (either elements or compounds). These data then can be used in analyzing chemical reactions.

HEAT ENGINE CYCLES

This section provides a brief description of several important heat engine cycles. A heat engine is a thermodynamic system that receives energy in the form of heat and that, in the performance of an energy transformation on a working fluid, does work. Heat engines function in cycles. An ideal heat engine works in accordance with the Carnot cycle (see chapter 4), while practical heat engines use thermodynamic cycles like the Rankine, Brayton, Stirling, or Otto cycles.

Engineers often categorize heat engines as being either external combustion (EC) engines or internal combustion (IC) engines. In an external combustion engine, the fuel is burned outside the engine. A steam engine operating on the Rankine cycle is an example. In an internal combustion engine, the fuel is burned inside the engine. A gasoline-fueled automobile engine operating on the Otto cycle is an example.

Rankine Cycle

The Rankine cycle is a fundamental thermodynamic heat engine cycle in which a working fluid is converted from a liquid to a vapor by the addition of heat (in a boiler), and the hot vapor is then used to spin a turbine, thereby providing a work output. Engineers often use this work output to rotate a generator and produce electricity. After passing through the turbine, the vapor is cooled in a condenser (rejecting heat to the surroundings). While being cooled, the working fluid changes from the vapor state to the liquid state; the liquefied working fluid then passes through a pump to start the cycle again. Developed in the 19th century by the Scottish engineer William John Macquorn Rankine, the Rankine cycle (in a variety of modern engineering configurations) is responsible for generating about 80 percent of the world's electric power.

The Tennessee Valley Authority's Shawnee Fossil Plant on the Ohio River near Paducah, Kentucky; the plant has 10 coal-fired generating units, which use the Rankine cycle to produce a total of about 9 billion kilowatt-hours of electricity each year. *(TVA)*

ORGANIC RANKINE CYCLE

In the organic Rankine cycle (ORC), the heat from a relatively low temperature source, such as the waste heat from a gas turbine engine exhaust, is transferred to an evaporator to boil an organic liquid into a vapor. The vaporized working fluid then expands through a turbine, spinning a generator to produce electricity. After the gaseous working fluid leaves the turbine, it passes through a condenser, becomes a liquid, and is pumped back into the evaporator.

The U.S. Department of Energy (DOE) is sponsoring research on the development of advanced, nonflammable working fluids that are suitable for use in organic Rankine cycle systems intended to harvest waste heat at industrial sites. Candidate organic fluids would operate at temperatures below 752°F (400°C). In principle, ORC heat engines can recover waste heat effectively from sources with temperatures as low as 158°F (70°C). Many candidate working fluids are flammable. One example is the aromatic hydrocarbon toluene ($C_6H_5CH_3$). If the evaporator develops a leak, direct contact with the input heat source could cause a potentially dangerous fire. Decomposition of the organic working fluid through lengthy operation of the ORC heat engine is another engineering concern. Decomposition by-products could cause fouling of the turbine. Successful research efforts involving advanced ORC heat engines will lead to improved recovery of the waste heat that accompanies many industrial operations.

Brayton Cycle

The Brayton cycle is the thermodynamic cycle developed by the American mechanical engineer George Brayton (1830–92). This cycle is used in all gas turbine engines. The working fluid remains a gas throughout the cycle. Brayton heat engines can operate either closed cycle or open cycle. In the closed cycle system, the gaseous working fluid is continuously recycled in a loop through the compressor, heat addition region, the power extracting turbine, and heat rejection region. In the open cycle system, the working fluid (typically air) passes just once through the system.

Stirling Cycle

The Stirling cycle is a thermodynamic cycle for a heat engine in which heat is added at constant volume, followed by isothermal expansion with heat addition. The heat is then rejected at constant volume, followed by isothermal compression with heat rejection. If a regenerator is used so that the heat rejected during the constant volume process is recovered during heat addition at constant volume, then the thermodynamic efficiency of the Stirling cycle approaches (in the limit) the efficiency of the Carnot cycle. The basic principle of the Stirling cycle heat engine was first proposed by the Scottish clergyman and inventor Robert Stirling (1790–1878).

In 1816, the Reverend Stirling invented the regenerator. The regenerator is a device used in a thermodynamic process for capturing and returning to the process heat that otherwise would be lost. The use of a regenerator helps increase the thermal efficiency of a heat engine cycle. Stirling obtained a patent for this "heat economizer" and then constructed the first practical prototype version of a Stirling engine two years later.

Aerospace engineers are exploring the use of candidate Stirling engine designs to provide electric power on future space missions. Similarly, mechanical engineers are studying Stirling engine applications in

An artist's rendering of a field of solar dish–Stirling engines in the southwestern United States; each solar dish generates electricity by focusing the Sun's rays onto a receiver, which transmits the heat to a Stirling engine. As envisioned here, each Stirling engine is a sealed system filled with hydrogen. As the hydrogen gas heats and cools, its pressure rises and falls. The change in pressure drives the pistons within the Stirling engine, producing work; this work drives a generator making electricity. *(DOE/SNLA)*

advanced solar-thermal systems for the generation of electricity. In 2008, the DOE reported on solar dish–Stirling engine projects that demonstrated a net solar-to-electric conversion reaching 30 percent. Each of the six demonstration units was capable of generating 25 kilowatts of daytime electric power. Encouraged by this research, energy officials suggest that a solar dish farm of Stirling engines, covering 11 square miles (28.5 km²) in the American Southwest, could (in principle) produce as much electricity per year as Hoover Dam. The average annual net generation for the Hoover Dam power plant is approximately 4.2 billion kilowatt-hours.

Otto Cycle

In 1861, the German inventor Nicolaus August Otto (1832–91) constructed a small experimental gasoline-powered engine that would eventually become the first practical alternative to the steam engine. Three years later, Otto joined with the German industrialist Eugen Langen (1833–95) and formed a company to manufacture internal combustion engines. By 1877, Otto had developed and patented the high-compression, four-stroke heat engine cycle that now bears his name. The four strokes associated with the ideal Otto cycle are the intake stroke, the compression stroke, the power stroke, and the exhaust stroke. Today, the Otto cycle serves as the operating principle for the great majority of the world's internal combustion engines.

The accompanying illustration shows the pressure-volume (p-V) diagram for an ideal Otto cycle heat engine. The discussion of the Otto cycle begins at stage 1. The start of the intake stroke corresponds to the end of the exhaust stroke. The pressure is near atmospheric pressure and the gas volume in the cylinder is at a minimum. During the intake stroke, the piston is pulled out of the cylinder with the intake valve open. From stage 1 to stage 2, the pressure remains constant; while the gas volume increases as a fuel/air mixture is drawn into the cylinder through the intake valve.

Stage 2 marks the beginning of the engine's compression stroke, as the intake valve closes. Between stage 2 and stage 3, the piston moves back into the cylinder, the gas volume decreases, and the gas pressure increases, as mechanical work (compression) is done on the gas by the piston. The combustion of the fuel/air mixture in the cylinder starts at stage 3. Since the combustion process occurs very quickly, engineers treat the process as a constant volume one. From stage 3 to stage 4, the heat released during combustion raises the temperature and pressure of the gaseous mixture.

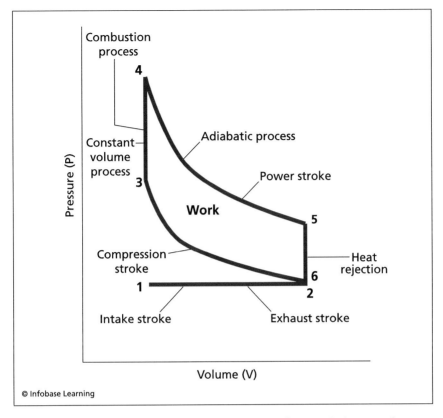

The pressure-volume (p-V) diagram for an ideal Otto cycle heat engine; engineers know that the work available each cycle for this four-stroke internal combustion engine equals the enclosed area (stages 3-4-5-6) of the p-V diagram. The engine's power equals the work per cycle times the number of cycles per second. *(adapted from NASA)*

The power stroke of the engine starts at stage 4. From stage 4 to stage 5, the piston is driven down the cylinder toward the crankshaft. The gas volume increases and the pressure decreases, as the combusted gas expands rapidly and does work on the piston. At stage 5, the engine's exhaust valve opens, and any residual heat in the combusted gas mixture is exchanged with the environment. From stage 5 to stage 6, the gas volume in the cylinder remains constant, as the pressure decreases back to atmospheric conditions.

Stage 6 (also called stage 2) represents the beginning of the engine's exhaust stroke. From stage 6 to stage 1, the piston moves back into the cylinder, the volume decreases, and the pressure remains constant. The thermodynamic cycle is now complete and ready to repeat itself.

RUDOLPH DIESEL

In the last decade of the 19th century, the German engineer Rudolph Christian Karl Diesel (1858–1913) began developing a new type of internal combustion engine. He patented this new compression-ignition heat engine in 1898 and then constructed a factory to manufacture them. Diesel became a millionaire as his engine gained popularity as a powerful alternative to the steam engines then being used in factories, railroad locomotives, and ships. The diesel engine also made the conventional (nonnuclear) submarine practical. Modern diesel engines are more robust and efficient than gasoline-powered Otto cycle engines. Consequently, diesel engines power many of world's trucks and certain model automobiles. Despite fame and fortune, Diesel's life came to a mysterious end on the evening of September 29, 1913. After dining onboard a mail steamer traveling between Belgium and Great Britain, Diesel went to his cabin and then simply vanished. A badly decomposed body with his personal effects was found 10 days later floating in the sea. Did he commit suicide or was he murdered? The reason for Diesel's apparent drowning death was never satisfactorily determined.

The Diesel cycle is similar to the Otto cycle in that engineers assume both cycles involve isentropic compression, as well as constant volume heat rejection of the exhaust gases. However, in the Diesel cycle heat addition takes place in a constant pressure process. This causes ignition and combustion to form part of the power stroke. The Diesel cycle describes the operating principle for practical compression-ignition heat engines. Under practical circumstances, a diesel engine can operate at much higher compression ratios than a spark-ignition (Otto cycle) engine. Spark-ignition engines often encounter detonation and preignition problems, when engineers attempt to compress fuel-air mixtures to very high pressures. Diesel engines do not experience such problems, because only air is compressed to high pressure and the fuel is not injected into

CHEMICAL ENERGY AND THERMODYNAMICS

Scientists define chemical energy as the energy stored in certain materials that is released by chemical reactions, such as the combustion of wood, coal, petroleum, or natural gas. Exothermic chemical reactions release energy in the form of heat, light, or acoustic waves (sound). Scientists denote an exothermic chemical reaction by a decrease in enthalpy ($\Delta H < 0$) and an increase in entropy ($\Delta S > 0$) of the system under study. During an exother-

the cylinder until conditions are appropriate for combustion. Diesel engines are used in applications where fairly constant engine speeds are desired, such as marine (ship) engines, locomotives, large trucks, and stationary electric generators. In contrast, spark-ignition (Otto cycle) engines provide the wide range of variable engine speeds favored in personal automobiles.

Air pollution–generating traffic snarls are one of the hallmarks of life in a highly mobile society. Diesel engines power many modern trucks and buses, while spark-ignition (Otto cycle) engines power most personal automobiles. *(EPA)*

mic reaction, heat flows from the reacting materials to the surroundings. Some chemical reactions may occur so rapidly as to be explosive.

The substances involved in an endothermic chemical reaction must absorb (gain) energy (typically in the form of heat from an external source) before the reaction can proceed. Scientists characterize an endothermic reaction by a positive flow of energy (heat) into the materials reacting and an increase in enthalpy ($\Delta H > 0$).

Thermochemistry is the branch of physical chemistry that deals with the heats of chemical reactions and the heats of formation of chemical compounds. The American theoretical physicist Josiah Willard Gibbs investigated the thermodynamics of chemical systems. His most important work *On the Equilibrium of Heterogeneous Substances* was

CHEMICAL EXPLOSIVES

Chemical explosives are substances that can experience very rapid exothermic chemical reactions (combustion), releasing heat and very hot gases that cause a sudden increase in pressure. These hot gases expand very rapidly, generating a blast (or shock) wave. The volume of hot gases produced by the chemical explosion is much greater than the original volume of the unexploded materials. Chemical bonding plays a major role in the behavior of explosive materials. Primary explosives detonate at once upon ignition; high explosives (such as TNT) typically only burn upon ignition but violently explode when detonated by a primary explosive.

Discovered in 1846, the pale yellow, oily liquid explosive called nitroglycerin ($C_3H_5[ONO_2]_3$) is shock sensitive and will detonate with the least vibration unless kept cold. In 1867, the Swedish industrialist and chemist Alfred Nobel (1833–96) tamed the fickle behavior of nitroglycerin by combining it with certain inert materials. Nobel patented his new explosive material and called it dynamite. Without a detonator, dynamite can be handled safely and is the chemical explosive frequently used in such civilian activities as mining, road building, and demolition. Because of this explosive, Nobel became very wealthy. After his death, his amassed fortune established the Nobel Prize. The Nobel Prize is an international award administered by the Nobel Foundation in Stockholm, Sweden. Every year since 1901, the Nobel Prize has been awarded for outstanding achievements in physics, chemistry, medicine or physiology, literature, and for peace. Each prize consists of a medal, personal diploma, and cash award. In 1968, the central bank of Sweden, Sveriges Riksbank, established the Sveriges Riksbank Prize in economic sciences in memory of Alfred Nobel.

The high explosive trinitrotoluene (TNT), $C_6H_2(NO_2)_3CH_3$, is a yellow-colored solid material that is convenient to handle because it is insensitive to shock and friction. TNT is one of the most commonly used military explosives. It also has

published between 1876–78 and introduced the foundational concept of chemical potential. The Gibbs function (G ≡ H − TS) describes the available (free) energy absorbed or released by a thermodynamic system as a result of a reversible constant temperature (T) and constant pressure process. Here, G represents the Gibbs function or Gibbs free energy, H

Called the Watusi Event, this large-scale chemical detonation experiment took place at the Nevada Test Site in September 2001 and produced the explosive yield equivalent of 40,000 pounds (18,140 kg) of trinitrotoluene (TNT). The blast was photographed from a mountaintop about 10 miles (16 km) from the detonation site. *(DOE/LANL)*

numerous industrial applications. The energy content of TNT is 1,977 Btu per pound-mass (4.6 megajoules per kilogram [MJ/kg]). Scientists and engineers have often used the energy content of TNT as a reference point with which to compare the performance of other chemical explosives or the performance of nuclear explosives. Explosive yields are usually expressed in terms of TNT equivalent. For example, gunpowder typically has an energy content of 1,290 Btu/lbm (3 MJ/kg); while dynamite contains about 3,229 Btu/lbm (7.5 MJ/kg).

represents enthalpy, and S represents entropy. Scientists find changes in the Gibbs function (ΔG) very useful in describing the conditions under which a chemical reaction will occur. For example, when the change in Gibbs free energy is positive ($\Delta G > 0$), then a chemical reaction can occur only if energy is added to force the thermochemical system away from its equilibrium position (that is, $\Delta G = 0$). If the change in Gibbs free energy is negative ($\Delta G < 0$), then the chemical reaction will proceed spontaneously to equilibrium. In 1901, the Royal Society of London awarded Gibbs the Copley Medal for his outstanding contributions to thermodynamics, especially his work regarding chemical potential and available (free) energy.

Chemical energy is liberated or absorbed during a chemical reaction. In such a reaction, energy losses or gains usually involve only the outermost electrons of the atoms or ions of the system undergoing change; here a chemical bond of some type is established or broken without disrupting the original atomic or ionic identities of the constituents.

Scientists discovered that the atoms in many solids are arranged in a regular, repetitive fashion and that the atoms in this structured array are held together by interatomic forces called chemical bonds. Chemical bonds play a significant role in determining the properties of a substance. There are several different types of chemical bonds. These include ionic bonds, covalent bonds, metallic bonds, and hydrogen bonds.

An ionic bond is created by the electrostatic attraction between ions that have opposite electric charges. Sodium chloride (NaCl), better known as table salt, is a common example of ionic bonding. Sodium (Na) is a silvery metal that has one valence electron to lose and form the cation (positive ion) Na^+. Chlorine (Cl) is a pale, yellow-green gas that has seven valence electrons and readily accepts another electron to form the anion (negative ion) Cl^-. The bond formed in creating sodium chloride molecule is simply electrostatic attraction between sodium and chlorine. When a large number of NaCl molecules gather together, they form an ionic solid that has a regular crystalline structure. In summary, the ionic bond discussed here is the result of the sodium atom transferring a valence electron to the valence shell of a chlorine atom, forming the ions: Na^+ and Cl^- in the process. Every Na^+ ion within the salt (NaCl) crystal is surrounded by six Cl^- ions, and every Cl^- ion by six Na^+ ions.

The second type of chemical bond is called the covalent bond. In the covalent bond, two atoms share outer-shell (valence) electrons. The molecular linkage takes place because the shared electrons are attracted to the positively charged nuclei (cores) of both atoms. One example is the hydrogen molecule (H_2), in which two hydrogen atoms are held tightly together. In 1916, the American chemist Gilbert Newton Lewis (1875–1946) proposed that the attractive force between two atoms in a molecule was the result of electron-pair bonding—more commonly called covalent bonding. During the following decade, developments in quantum mechanics allowed other scientists to quantitatively explain covalent bonding. In most molecules, the atoms are linked by covalent bonds. Generally, in most molecules hydrogen forms one covalent bond; oxygen forms two covalent bonds; nitrogen three covalent bonds; and carbon four covalent bonds. In the water molecule, the oxygen atom bonds with two hydrogen atoms. The H—O bonds formed as the oxygen atom shares two pairs of electrons; while each hydrogen atom shares only one pair of electrons, which scientists refer to as polar covalent bonds.

In metallic bonding, electrons are freely distributed so that many metal atoms share them. Immersed in this sea of negative electrons, the positive metal ions form a regular crystalline structure. Consider the metal sodium as an example. This metal is made up of individual sodium atoms, each of which has 11 electrons. In metallic bonding, every sodium atom releases one valence electron, which then moves throughout the metallic crystal attracted to the positive Na^+ ions. It is this attraction that holds the metallic crystal together.

Finally, hydrogen bonding is a weak to moderate attractive force due to polarization phenomena. This type of bonding occurs when a hydrogen atom that is covalently bonded in one molecule is at the same time attracted to a nonmetal atom in a neighboring atom. Hydrogen bonding typically involves small atoms characterized by high electronegativity, such as oxygen, nitrogen, or fluorine. Chemists define electronegativity as the attraction of an atom in a compound for a pair of shared electrons in a chemical bond. Under certain conditions during physical state transitions, hydrogen bonding occurs in the water molecule. Hydrogen bonds are also common in most biological substances.

CHAPTER 6

Harvesting Energy Locked in Fossil Fuels

This chapter discusses the three principal fossil fuels: coal, petroleum (oil), and natural gas. According to the *2010 Electric Power Annual* published by the U.S. Energy Information Administration (EIA), 48 percent of the nation's 4.1 trillion kilowatt-hours of electricity came from coal; 21 percent from natural gas; and 1 percent from petroleum. Along with uranium (discussed in chapter 8), coal, petroleum, and natural gas are also categorized as nonrenewable energy resources because they cannot be replenished in a short period of time. The fossil fuels formed during the Carboniferous period, which took place between 360 and 286 million years ago during the Paleozoic era. At the time, the land was covered with swamps populated by giant trees, tall ferns, and other large leafy plants. The oceans, seas, and lakes were filled with millions of very tiny plants called green algae. Geologists suggest that some coal deposits originate from the Cretaceous period—the time of the dinosaurs.

COAL

Coal, the rock that burns, represents one of the world's major energy resources. Coal is the chemical fuel that provides the heat necessary to generate more than half of electricity used in the United States.

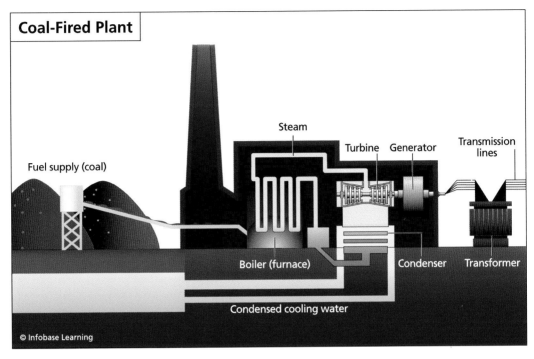

Coal-Fired Plant

Steam

Turbine Generator

Transmission lines

Fuel supply (coal)

Boiler (furnace)

Condenser Transformer

Condensed cooling water

© Infobase Learning

Schematic of a modern coal-fired plant to generate electricity *(TVA)*

Coal forms from decayed and compacted plant remains. The first stage of coal formation is the creation of peat, which is partially decayed plant material that is protected from further oxidation (decay) by submersion in water. Over time, as peat is subjected to increasing lithostatic pressure and temperature, it gradually transforms under these geologic forces into coal in the following sequence: lignite, subbituminous coal, bituminous coal, and finally anthracite.

Lignite forms first, as increased pressures and heat within Earth's crust cause buried peat to dry and harden. It is a brown-colored coal with high moisture and ash content and relatively low heating value (about 4,000–8,300 Btu/lbm [9.3–19.3 MJ/kg]). Lignite has a relatively low carbon content, which ranges from about 25 to 35 percent by weight.

As the geologic pressure increases, lignite transforms into the next type of coal. Subbituminous coal is dull black in color and has a carbon content ranging from 35 to 45 percent. This type of coal has a heating value ranging from 8,300 to 13,000 Btu/lbm (19.3–30.2 MJ/kg). Although subbituminous coal has a lower heating value than bituminous coal or anthracite, utilities often prefer to burn it in their electric power–generating stations because of its generally lower sulfur content.

At even greater geologic pressure, subbituminous coal transforms into bituminous coal. Also known as soft coal, bituminous coal is the most plentiful type of coal found in the United States. This type of coal is used primarily to generate electricity and to manufacture coke for the steel industry. Coke is the porous but hard gray-colored carbon fuel that results when bituminous coal is heated to high temperatures under tightly controlled air flow conditions intended to drive off easily vaporized contaminants. Bituminous coal has a carbon content that ranges from 45 to 86 percent and a heating value that ranges from about 10,500 to 15,000 Btu/lb (24.4–34.9 MJ/kg).

Sometimes called hard coal, anthracite forms when tectonic processes place bituminous coal under conditions of great geological pressure, such as when mountain ranges form. Anthracite coal has the highest carbon content of all coals (up to 98 percent) and the highest heat value (about 15,000 Btu/lb [34.9 MJ/kg]). In the United States, anthracite coal deposits are found only in the northeastern portion of Pennsylvania. This form of coal is used for the generation of electric power and also for heating homes.

CARBON CYCLE

Carbon is the basis for life on Earth. This essential element moves through the planet's biosphere in a great act of natural recycling called the global carbon cycle. Scientists find it helpful to divide the global carbon cycle into two components: the geological carbon cycle, which operates over very long time periods (millions of years); and the physical/biological carbon cycle, which takes place over much shorter periods (days to millennia).

The carbon that now cycles through Earth's various planetary systems (that is, the biosphere, atmosphere, hydrosphere, and lithosphere) was present in the ancient solar nebula from which the solar system began to form about 5 billion years ago. As Earth emerged as a distinct celestial object about 4.6 billion years ago, the young planet's surface was extensively bombarded by carbon-bearing meteorites. Over time, these meteorite impacts steadily increased the planet's carbon content. Since the end of the period of great cosmic collisions, carbonic acid has slowly and steadily combined with calcium and magnesium in Earth's crust to form

CARBON SEQUESTRATION

Energy demand projections by the DOE (in 2010) indicate that fossil fuels will continue to serve as the major source of energy throughout the world for many decades. Since the consumption of fossil fuels remains intimately linked to national security and economic vitality, scientists remain very busy investigating ways of keeping the atmospheric concentrations of carbon dioxide from rising. One approach to controlling carbon emissions from fossil fuels is called carbon sequestration.

Scientists define carbon sequestration as the capture and long-term storage of carbon dioxide and other greenhouse gases (such as methane) that would otherwise enter Earth's atmosphere. They suggest the greenhouse gases can either be captured at the point of origin (direct sequestration) or else removed from the atmosphere (indirect sequestration). The captured carbon dioxide can then be stored in underground reservoirs (geological sequestration), injected into deep portions of the oceans (ocean sequestration), or stored in vegetation and soils (terrestrial sequestration).

Geological sequestration involves the storage of captured carbon dioxide in depleted oil and gas reservoirs, in coal seams that can no longer be practically mined, or possibly within underground saline formations. The high pressure injection of carbon dioxide into depleted oil and gas reservoirs may also force any remaining oil or gas toward production wells—facilitating enhanced hydrocarbon recovery operations. Ocean sequestration involves directly injecting carbon dioxide deep into the ocean. Although carbon dioxide is soluble in seawater and the oceans naturally absorb and release huge amounts of carbon dioxide, this proposed technique is not without controversy. The controversy centers around what impact deepwater injection activities might have on the ocean and various marine ecosystems. Terrestrial sequestration involves removal of carbon dioxide from the atmosphere by means of vegetation and soils. Ecosystems that offer significant opportunities for enhanced carbon sequestration include forests, biomass crops, grasslands, and peat lands.

Advanced sequestration concepts, like converting captured greenhouse gases into rocklike solid materials, are also being investigated. Scientists are examining the feasibility of using minerals—such as magnesium carbonate—for carbon capture and storage. Before any carbon sequestration approach is implemented on a large scale this century, scientists and engineers must demonstrate that the selected approach is technically practical, economic, and environmentally acceptable.

carbon-containing chemical compounds called carbonates. Carbonic acid (H_2CO_3) is a weak acid that forms when gaseous carbon dioxide (CO_2) dissolves in water (H_2O). Through weathering, the carbonic acid combined with the calcium and magnesium found in Earth's crust to form chemical compounds, such as calcium carbonate (limestone). This process was continual but very gradual. Another natural process called erosion washed the carbonates into the ancient oceans. Once in the ocean, these carbon-bearing compounds precipitated out of the ocean water and formed layers of sediment on the ocean floor.

The process of plate tectonics then pushed these carbon-bearing sediments underneath the continents. Geologists refer to this activity as subduction. Once deep within the lithosphere, the limestone and other carbon-bearing sediments experienced increased heat and pressure. The carbonates melted, reacted with other minerals, and released carbon

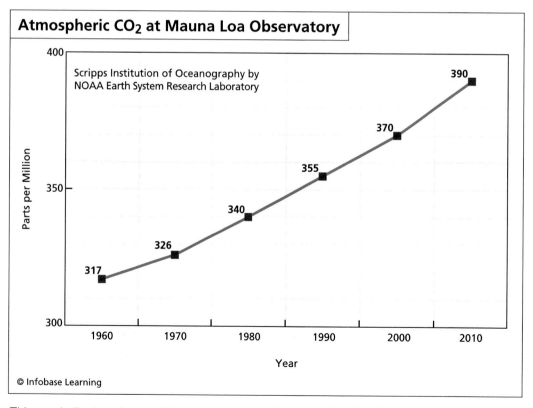

This graph displays the monthly mean atmospheric carbon dioxide (CO_2) data (in parts per million) recorded at the Mauna Loa Observatory, Hawaii, from March 1958 to July 2010. *(NOAA)*

dioxide. Volcanic eruptions then returned the released carbon dioxide to Earth's atmosphere. Scientific evidence suggests that there is a natural balance in the geological carbon cycle between weathering, erosion, subduction, and volcanism. The geological carbon cycle still regulates atmospheric carbon dioxide concentrations—but operates over time periods of hundreds of millions of years.

In the physical/biological carbon cycle, living things play a major role in moving carbon through the biosphere. During the process of photosynthesis, green plants absorb carbon dioxide from Earth's atmosphere and use sunlight (energy) to create fuel (glucose and other sugars) for constructing more complex plant structures. Scientists describe photosynthesis as the natural process through which chlorophyll-bearing green plants use the energy content of sunlight to make carbohydrates from atmospheric carbon dioxide and water. In the basic process, plants transform input hydrogen, oxygen, and carbon into the stable organic compound glucose ($C_6H_{12}O_6$) and release water and oxygen back into Earth's environment.

Chemists define carbohydrates as any member of a large class of carbon-bearing compounds—including sugars, starches, cellulose, and similar compounds. Carbohydrates (sugars) serve as the enabling fuel for living things and allow the organisms to grow and reproduce. Here on Earth, plants and animals metabolize (burn) carbohydrates and other nutrient molecules. Through aerobic respiration (the opposite of photosynthesis) living organisms release the energy stored in carbohydrates by combining nutrient organic molecules (such as glucose) with oxygen and producing water and carbon dioxide. The released carbon dioxide carries carbon back into the atmosphere. Another natural process, called decomposition (the digestion of dead or decaying organic matter by bacteria and fungi), also returns carbon (fixed by photosynthesis) back into the atmosphere. Carbon circulates through the biosphere because of the linkage between photosynthesis, respiration, and decomposition.

Photosynthesis and respiration play important roles in moving carbon through the biosphere—on timescales far shorter than those involved in the geological carbon cycle. The biological processes for various living organisms are elegantly complex; only the very basic carbon exchange activities are mentioned here. Scientists currently estimate that the yearly quantity of carbon fixed by photosynthesis and released back to the atmosphere by respiration is about 1,000 times greater than the amount of carbon that moves through the geological carbon cycle on an annual basis.

Millions of years ago, some of the carbon involved in biological processes was not released back into the atmosphere as carbon dioxide. Instead, buried deposits of dead plants (on land) and certain marine lifeforms (in the oceans) became compressed over time by layers of sediment, eventually forming fossil fuels (such as coal, oil, and natural gas). The carbon locked in these fossil fuels remained trapped within Earth's crust for millions of years—until humans mined the various fuels and began burning them.

Since the start of the Industrial Revolution in the 17th century, the carbon dioxide content of Earth's atmosphere has increased from about 280 parts per million (ppm) to a current day (2010) value of about 390 ppm. Human activities—especially the consumption of enormous quantities of fossil fuels and large-scale deforestation—now exert a measurable influence on the planet's global carbon cycle. Many scientists are alarmed at the increased quantity of carbon dioxide in the atmosphere and warn about the dire consequences of global warming due to greenhouse gas buildup. Climate models indicate that increased greenhouse gas concentrations (including methane [CH_4] and carbon dioxide) have been the primary driver of Earth's increasing surface temperatures. Improved ways of tracing the amount and pathways of carbon as this element travels throughout Earth's biosphere are important aspects of Earth system science. Such scientific activities should provide data that lead to a more responsible level of stewardship of humans' home planet in this century.

PETROLEUM

The word *petroleum* is a combination of the ancient Greek words for rock and oil. Known to many ancient peoples because of seepage, crude oil is a naturally occurring liquid found in various underground geologic formations. Typically black or dark brown crude oil consists of a complex collection of hydrocarbons, ranging from C_5H_{12} to $C_{18}H_{38}$ in molecular composition. Engineers regard petroleum as an important primary energy source and call it a fossil fuel because of its prehistoric origins.

It is very common in daily language to interchange the words petroleum and oil. Almost everyone does it. Material scientists define oil as a substance that near room temperature (68°F [20°C]) exhibits a viscous liquid state. Typically, an oily liquid does not mix with water. Scientists use the words immiscible (non-mixing) and hydrophobic to describe these characteristics. However, one type of oil is usually tolerant of other types of oil and will usually mix well with them. Scientists often use the term lipophilic (literally fat-loving) to describe this tendency. Chemists treat a liquid as lipophilic if it is able to dissolve much more easily in other oily organic compounds than in water. The word oil is actually a rather nonscientific term. People encounter all types of oil in daily life. There are cooking oils (like peanut oil), animal fat oils, salad oils, fuel oils, body lubricants (like suntan oil), and even fragrant holy oils (chrisms). In this chapter, the word oil means petroleum, unless specified otherwise.

Petroleum and natural gas formed from the remains of ancient marine plants and animals that died millions of years before the dinosaurs roamed Earth. When these tiny prehistoric marine creatures and plants died, their organic remains settled on the ocean floor. (Scientists now speculate that coal formed mostly from the decay of ancient plants, while petroleum formed mostly from the decay of ancient animals, especially from their fat.) Over time, sand and silt covered the decaying material.

A petroleum-producing well in Osage County, Oklahoma *(DOE)*

As millions of years passed, the remains sank deeper into Earth's crust. Deep within the crust, the intense heat and pressure slowly converted the ancient organic remains into crude oil and natural gas. Methane (CH_4) and ethane (C_2H_6) are examples of natural gas. Crude oil is a smelly, light brown-to-black viscous liquid that has typically accumulated in underground pockets called reservoirs.

Geologists explore for oil reservoirs by studying rock samples taken from suspected oil-bearing regions. If the underground site appears promising, drilling operations begin. On land (or at sea), engineers employ a structure called a derrick to house the pipes and tools going into the well. When finished, a successful, properly drilled well brings a steady flow of crude oil to the surface.

Prior to the advent of modern drilling technologies and pressure control equipment in the early decades of 20th century, the rapid release of crude oil from an underground reservoir would often accompany a successful drilling operation. Oil industry workers coined the iconic expression gusher to describe the wild situation, when it quite literally "rained oil" all around the drilling site. The precipitation of crude oil continued until workers could successfully cap the new well.

Even today, a blowout can take place when the pressure control equipment at an oil well fails. On April 20, 2010, the largest underwater blowout in history took place in the Gulf of Mexico off the coast of Louisiana. A violent explosion at a mobile offshore drilling platform named *Deepwater Horizon* killed 11 workers, and the platform burned uncontrollably until it sank. For the next three months, crude oil spewed from the damaged wellhead on the seafloor, threatening marine life and the ecologically fragile coastal regions of Louisiana, Mississippi, Alabama, and Florida. The multinational oil company British Petroleum was leasing the drilling platform. On July 15, emergency response workers managed to achieve a static kill on top of the well by pumping mud and cement into the well's containment cap. American officials overseeing response to this major environmental disaster estimated that a total of about 5 million barrels (205 million gallons) of oil had spilled from the ruptured well into the Gulf of Mexico before the capping operations stopped the leakage. By mid-September, a relief well successfully intersected the blown-out oil well some 18,000 feet (5,500 m) below the Gulf of Mexico's seabed. This allowed petroleum engineers to permanently seal the rupture from below. They used the relief well to perform a companion bottom kill by injecting mud and cement. The five-month saga represents the world's worst accidental oil spill.

Fireboats battling the blazing remains of the *Deepwater Horizon* mobile offshore oil drilling platform on April 21, 2010. The explosion killed 11 workers and sent millions of gallons of crude oil into the Gulf of Mexico—threatening ecologically fragile coastal regions of Louisiana, Mississippi, Alabama, and Florida. *(USCG)*

Workers extract crude oil from a normally operating well and ship it to a refinery for processing. Two major techniques for shipping oil are by sea (using tankers or barges) and overland (using a network of pipelines). Both techniques usually transport large quantities of crude oil efficiently each day. However, accidents can occur and result in the spill of crude oil into environmentally sensitive regions. For example, on March 24, 1989, an oil tanker, the *Exxon Valdez,* hit a reef in Prince William Sound, Alaska, and spilled an estimated 10.8 million gallons (40.9×10^6 L) of crude oil. Regarded as a one of the major human-caused environmental disasters, the oil contaminated some 1,300 miles (2,090 km) of ecologically fragile coastline and covered about 11,000 square miles (28,510 km^2) of the Pacific Ocean.

The 10 leading crude oil–producing countries (in descending order) are: Saudi Arabia, Russia, United States, Iran, China, Canada, Mexico, United Arab Emirates, Kuwait, and Venezuela. The top crude oil–producing states in the United States are Texas, Alaska, California, Louisiana, and New Mexico. More than one-fourth of the crude oil produced in the United States is extracted offshore in the Gulf of Mexico. Crude oil is traded in barrels (bbls). Due to density differences after refining and processing, a 42-gallon (159 L) barrel of crude oil typically yields about 45 gallons (170 L) of refined petroleum products.

PETROLEUM REFINING AND PROCESSING

A modern oil refinery is an amazing factory that takes in crude oil and converts it into a variety of more useful products, including transportation fuels, heating oil, fertilizers, petrochemicals, asphalt, and many other materials. A modern refinery operates 24 hours a day every day of the year. Generally, a refinery will convert a 42-gallon (159-L) barrel of crude oil into 18.56 gallons (70.25 L) of gasoline, 10.31 gallons (39.0 L) of diesel fuel, 1.68 gallons (6.36 L) of heavy fuel oil (residuum), 4.07 gallons (15.41 L) of jet fuel, 1.38 gallons (5.22 L) of other distillates (such as heating oil), 1.72 gallons (6.51 L) of liquefied petroleum gases (LPG), and 7.01 gallons (26.53 L) of other products. These other products include ink, crayons, bubble gum, dishwashing liquids, deodorant, eyeglasses, tires, CDs and DVDs, heart valves, and ammonia.

Each refinery is a unique assembly of processing equipment. However, every petroleum refinery performs three basic steps: separation, conversion, and treatment. All refineries have distillations units that separate crude oil by boiling point. Heavy petroleum fractions are on the bottom of the distillation tower, while light fractions occupy the top. This arrangement allows for the separation of various petrochemicals.

Chemists describe petroleum refining as the conversion of the relatively abundant alkanes (found in crude oil) into unsaturated hydrocarbons, aromatic hydrocarbons, and hydrocarbon derivatives. Alkanes (paraffins) are saturated hydrocarbons with the general molecular formula C_nH_{2n+2}. Methane (CH_4), hexane ($CH_3[CH_2]_4CH_3$), and octane ($CH_3[CH_2]_6CH_3$) are examples of straight-chain alkanes. In the fractional distillation of petroleum, chemist engineers vaporize the input crude oil and then column separate the components according to boiling points. The lower boiling point constituents (like naphtha and kerosene) reach the upper portions of the distillation tower, while the higher boiling point components remain in the lower parts of the tower. Chemical engineers define naphtha as the lightest and most volatile fraction of the liquid hydrocarbons in a distillation tower. They use naphta as the feedstock in the production of high-octane gasoline. The reformer converts naphthas (that have low octane ratings) into gasolines with higher octane ratings.

Gasoline is a hydrocarbon fuel used as the chemical energy source for many internal combustion engines, especially those found in automobiles. Commercial grades of gasoline have molecular compositions ranging from C_5H_{12} to $C_{12}H_{26}$ and boiling points from 104°F to 392°F (40°C to 200°C). On occasion, the air-gasoline mixture in an internal combustion engine prematurely ignites before the spark plug fires. Automotive engineers refer to this undesirable phenomenon as knocking. In the 1920s, chemical engineers created an arbitrary scale for automotive engine performance based on a gasoline mixture's ability to prevent knocking. They discovered that certain blends caused less knocking and assigned these hydrocarbon blends a higher octane rating. Chemical engineers perform a variety of processes and treatments in a modern refinery to produce high octane gasoline and other desired petroleum products.

They use the process of cracking to break down larger hydrocarbon molecules into smaller sized ones. The cracking process allows chemical engineers and chemists to modify natural materials into a variety of different new ones, more suitable for human needs. In addition to fuels for transportation and heating, crude oil yields a wide variety of petrochemicals that form an integral part of modern living.

The process of alkylation allows chemical engineers to combine certain small hydrocarbon molecules (below the molecular size range of gasoline) into the large hydrocarbon molecules typically found in gasoline. For example, by reacting isobutylene (C_4H_8) with propane (C_3H_8), the engineers formed highly branched product molecules that were just the proper size for gasoline and were also high in octane number. The hydrocarbon molecule octane has the chemical formula, $CH_3(CH_2)_6CH_3$.

Finally, the coker in a petroleum refinery uses heat to transform a portion of the higher molecular weight residuum (residual crude oil) in the distillation tower into lower molecular weight hydrocarbons, such as liquid petroleum gases (LPGs) and gasoline. The remainder of the residuum becomes industrial fuel or asphalt base. This has been a very brief summary of the complex and elegant processing technologies used by chemical engineers in modern petroleum refineries to provide a wide variety of contemporary products that include automotive fuels, pharmaceuticals, plastics, perfumes, and pesticides.

NATURAL GAS

Natural gas is mostly a mixture of methane (CH_4), ethane (C_2H_6), and propane (C_3H_8), with methane making up about 73 to 95 percent of the total. Often found during drilling for petroleum, natural gas was once treated as a nuisance and burned off. Today, natural gas is mostly injected back into the ground for later recovery and use. When injected under pressure back into an oil formation, natural gas can also stimulate petroleum recovery and production.

The main ingredient in natural gas is methane, a gas (or compound) composed of one carbon atom and four hydrogen atoms. Millions of years ago, the remains of plants and animals (diatoms) decayed and built up in thick layers. Scientists call this decayed matter from plants and animals organic material since it was once alive. Over time, the sand and silt changed to rock, covered the organic material, and trapped it beneath the rock. Pressure and heat changed some of this organic material into coal, some into oil (petroleum), and some into natural gas—tiny bubbles of odorless gas. In some places, gas escapes from small gaps in the rocks into the air. Then, if there is enough activation energy from lightning or a fire, it burns. When people first saw the flames, they experimented with them and learned they could use them for heat and light.

In about 200 B.C.E., the Chinese used natural gas to assist in the production of salt from salt brine. French explorers traveling in North America in 1616 encountered Native Americans who were burning natural gas as it seeped into and around Lake Erie. The city of Baltimore used natural gas to light streetlights as early as 1816. An American businessperson named William Hart dug the first successful well intended to produce natural gas in 1821 in the town of Fredonia, New York. In 1859, another American businessperson named Edwin Drake drilled the first oil well in the United States in the village of Titusville, Pennsylvania. He discovered both petroleum and natural gas at a depth of just 69 feet (21 m) below the surface. Science historians generally regard Drake's well as the start of the American petroleum and natural gas industries.

Today, the search for natural gas begins with geologists who study the structure and processes of the Earth. They locate the types of rock that are likely to contain gas and oil deposits. The tools used by geologists include

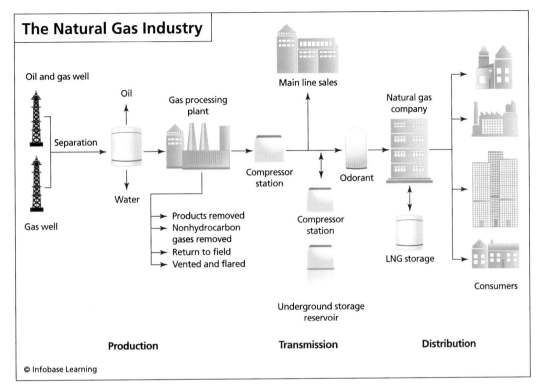

The Natural Gas Industry

Oil and gas well

Oil

Gas processing plant

Main line sales

Natural gas company

Separation

Water

Compressor station

Odorant

Gas well

→ Products removed
→ Nonhydrocarbon gases removed
→ Return to field
→ Vented and flared

Compressor station

LNG storage

Consumers

Underground storage reservoir

Production

Transmission

Distribution

© Infobase Learning

This diagram shows the major components of the natural gas industry in the United States. (DOE)

seismic surveys that are used to find the right places to drill wells. Seismic surveys use echoes from a vibration source at the Earth's surface (usually a vibrating pad under a truck built for this purpose) to collect information about the rocks beneath. Sometimes it is necessary to use small amounts of dynamite to provide the vibration that is needed.

Scientists and engineers explore a chosen area by studying rock samples from deep below the ground and taking measurements. If the site seems promising, drilling begins. Some of these areas are on land, but many are offshore, deep in the ocean. Once the gas is found, it flows up through the well to the surface of the ground and into large pipelines.

Some of the gases that are produced along with methane, such as butane (C_4H_{10}) and propane (also known as by-products), are separated and cleaned at a gas-processing plant. The by-products, once removed, are used in a number of ways. For example, propane can be used for cooking on gas grills or to operate portable electric power generators. Dry natural gas is called consumer-grade natural gas. In addition to natural gas production, the U.S. gas supply is augmented by imports, withdrawals from storage, and by supplemental gaseous fuels. Most of the natural gas consumed in the United States is produced in the United States. Some is imported from Canada and shipped to the United States in pipelines. Increasingly, natural gas is also being shipped to the United States as liquefied natural gas (LNG).

People can also use machines called digesters that turn today's organic material (plants, animal wastes, etc.) into natural gas. This process replaces waiting for millions of years for the gas to form naturally.

Natural gas is moved by pipelines from the producing fields to consumers. Because natural gas demand is greater in the winter, it is stored along the way in large underground storage systems, such as old oil and gas wells or caverns formed in old salt beds. The gas remains there until it is added back into the pipeline when people begin to use more gas, such as in the winter to heat homes.

When the gas gets to the communities where it will be used (usually through large pipelines), it flows into smaller pipelines that engineers call mains. Very small lines, called services, connect to the mains and go directly to the homes and buildings where it will be used. When chilled to very cold temperatures, approximately –260°F (–162°C), natural gas changes into a liquid and can be stored in this form. Because it takes up only 1/600th of the space that it would in its gaseous state, LNG can be loaded onto tankers (large ships with several domed tanks) and moved across the ocean to other countries. When this LNG is received in the United States, it can be shipped by truck to be held in large chilled tanks close to users or turned back into gas when it is ready to put in the pipelines.

LNG is natural gas that has been cooled to about –260°F (–162°C) for shipment and/or storage as a liquid. The volume of the liquid is about 600 times smaller than in its gaseous form. In this compact form, natural gas can be shipped in special tankers to receiving terminals in the United States and other importing countries. At these terminals, the LNG is returned to a gaseous form and transported by pipeline to distribution companies, industrial consumers, and power plants.

Liquefying natural gas provides a means of moving it long distances where pipeline transport is not feasible, allowing access to natural gas from regions with vast production potential that are too distant from end-use markets to be connected by pipeline. Representing a possible vision of tomorrow's surface transportation system in many parts of the

A municipal bus fueled by compressed natural gas *(EPA)*

United States, experimental vehicles, including buses, are being operated on either compressed natural gas (CNG) or LNG fuel.

According to the DOE, about 25 percent of energy used in the United States in 2009 came from natural gas. That year, Americans consumed 22.84 trillion cubic feet (Tcf) (about 0.65 trillion m³) of natural gas. Natural gas is used to produce steel, glass, paper, clothing, brick, electricity and as an essential raw material for many common products. Some products that use natural gas as a raw material are paints, fertilizer, plastics, antifreeze, dyes, photographic film, medicines, and explosives. Slightly more than half of the homes in the United States use natural gas as their main heating fuel. Natural gas is also used in homes to fuel stoves, water heaters, clothes dryers, and other household appliances.

Natural gas burns more cleanly than other fossil fuels. It has fewer emissions of sulfur, carbon, and nitrogen than coal or oil, and when it is burned, it leaves almost no ash particles. Being a cleaner fuel is one reason that the use of natural gas, especially for electricity generation, has grown so much. However, as with other fossil fuels, burning natural gas produces carbon dioxide, which is a greenhouse gas.

Similar to other fuels, natural gas also affects the environment when it is produced, stored, and transported. Because natural gas is made up mostly of methane (another greenhouse gas), small amounts of methane can sometimes leak into the atmosphere from wells, storage tanks, and pipelines. The natural gas industry is working to prevent any methane from escaping.

Exploring and drilling for natural gas will always have some impact on land and marine habitats. But new technologies have greatly reduced the number and size of areas disturbed by drilling, sometimes called footprints. Furthermore, engineers now use horizontal and directional drilling techniques, which make it possible for a single well to produce gas from much bigger areas than in the past.

Natural gas pipelines and storage facilities have a good safety record. This is important, because when natural gas leaks it can cause explosions. Since raw natural gas has no odor, natural gas companies add a distinctive, smelly substance to it—so that people will know if there is a leak. A person with a natural gas stove may have smelled this rotten egg smell of natural gas when the pilot light has gone out.

CHAPTER 7

Manipulating Matter's Electromagnetic Properties

This chapter discusses electromagnetism, one of the most interesting and important energetic properties of matter. From telecommunications to computers, from electronic money transfers to home entertainment centers, the application of electromagnetic phenomena dominates contemporary civilization. People everywhere are immersed in a world controlled by the manipulation and flow of electrons.

MAGNETISM AND THE MYSTERIES OF THE LODESTONE

The history of magnetism extends back to antiquity. History suggests that the Greek playwright Euripides (ca. 480–406 B.C.E.) applied the word *magnes* (μαγνης) to a type of ore (now called magnetite) that displayed a preferential attraction for iron. The English word *magnet* traces its origin to the ancient Greek expression "stone of magnesia" (μαγνητης λιθος). The reference is to an interesting type of iron ore, named a lodestone by alchemists, that is found in Magnesia, a region in east-central Greece. Scientists also named the chemical elements magnesium (Mg) and manganese (Mn) for substances associated with this geographic location.

Lodestone is an intensely magnetic form of magnetite—a mineral consisting of iron oxide (Fe_3O_4). Although magnetite is a commonly found mineral, lodestone is rare. Contemporary scientific research suggests that not all pieces of magnetite can become a lodestone; a certain crystal structure and composition that experiences a strong transient magnetic field is required. Nature provides this transient magnetic field during a lightning strike. Lightning involves a large electric current lasting just a fraction of a second, during which time the enormous electric charge also produces a strong magnetic field. Experiments have demonstrated how lightning strikes could transform mineral samples of magnetite with the appropriate crystalline structure into lodestones.

Until the Chinese used lodestone to develop an early mariner's compass in the latter portion of the 11th century, little progress was made by human beings in either understanding or applying the phenomenon of natural magnetism. The notion of a compass, consisting of a slender sliver of lodestone suspended on a string, appeared in Europe about a century later. The compass is a device that uses the intrinsic properties of a magnetized metal pointer (such as lodestone) to indicate the general direction of Earth's magnetic North Pole. Earth is actually a gigantic magnet with

Lightning streaks across the night sky. On average, Earth's surface experiences about 100 cloud-to-ground lightning strikes every second. Each bolt may contain up to 1 billion volts of electricity. *(NPS)*

one pole geographically located in northern Canada (called the north magnetic pole) and the other pole in Antarctica (south magnetic pole). The North Pole–pointing mariner's compass improved the art of open-water navigation and greatly facilitated global exploration by sailing ships of European nations in the 15th and 16th centuries.

In 1600, the British physician and scientist William Gilbert (1544–1603) published *De Magnete (On the Magnet),* the first technical book on the subject. His pioneering study distinguished between electrostatic and magnetic effects. Gilbert received an appointment as the royal physician to Great Britain's Queen Elizabeth I in 1601. He was also an excellent scientist, who explored all previous observations concerning the magnetic and electrostatic properties of matter. As part of the ongoing Scientific Revolution, Gilbert systematically investigated and extended these ancient observations and frequently enhanced them with clever experiments of his own construction.

Testing many different materials, he listed all that exhibited the same electrostatic property as amber. In so doing, he introduced the Latin word *electricus* to describe an amberlike attractive property. The ancient Greek word for amber is electron (ηλεκτρον) from which the modern English words *electron* and *electricity* derive. Gilbert also examined natural magnetism very thoroughly and demonstrated that he could magnetize steel rods by stroking them with lodestone.

Gilbert's investigations of lodestone revealed many other important properties of natural magnets. His research led him to the very important conclusion that Earth is a giant magnet, thereby explaining the mysterious operation of the mariner's compass. He noted that every magnet has two poles—now arbitrarily called a north (or source) pole and a south (or sink) pole. As a result of his lodestone experiments, Gilbert discovered that like magnetic poles repel each other; while unlike magnetic poles attract each other. His experiments showed that north-south poles attract, while north-north or south-south poles repel.

The next major breakthrough in the physics of magnetism occurred quite serendipitously. At the University of Copenhagen, the Danish physicist Hans Christian Ørsted (1777–1851) was investigating what happens when an electric current flows through a wire. On April 21, 1820, he made preparations for laboratory demonstrations about electricity and about magnetism. He used the newly invented battery (Voltaic pile) as his source of electricity. As Ørsted moved his equipment around, he noticed that a current-carrying wire caused an unexpected deflection in the needle of a nearby compass. By chance, that particular wire was carrying a current, because he was also planning to discuss how the flow electricity from a Voltaic pile could cause heating in a wire. The deflection of the nearby compass fortuitously caught his attention, and the event changed physics.

Over the next several months, Ørsted repeated the experiment and watched the compass needle deflect whenever current flowed through a nearby wire. He could not explain how the flow of electricity from a battery could cause this magnetic effect, but, as a scientist, he felt obliged to report his unusual observations to Europe's scientific community. Ørsted's discovery had clearly linked electricity and magnetism for the very first time. The discovery immediately encouraged other scientists, such as the French physicist André-Marie Ampère, to launch their own comprehensive investigations of electromagnetism.

After Gilbert published *De Magnete,* other researchers turned their attention to magnetism. Some were fascinated by the fact that, if they cut a bar magnet in half, each half suddenly developed its own magnetic north and south poles. With this discovery, a more tantalizing technical question arose: Could scientists isolate a magnetic monopole—that is, a hypothetical subnuclear particle functioning as just a north pole or a south pole magnet? Up through the end of the 20th century, scientists searched for evidence of a magnetic monopole, but failed. The contemporary consensus within today's scientific community is that nature does not support isolated magnetic poles. The detailed answer as to why this appears so lies in quantum mechanics and the characteristics of the magnetic fields associated with spinning electrons—specifically, the behavior of a material's magnetic dipole moments. Yet, some scientists working at large particle accelerator facilities continue their quest for an elusive, hypothetical particle that functions as a quantum of magnetism.

Each of an atom's electrons has an orbital magnetic dipole moment and a spin magnetic dipole moment. The material exhibits magnetic properties when the combination of each of these individual dipole moments produces a net magnetic field. Although a detailed, quantum-mechanical description of the phenomena that give rise to magnetism is beyond the scope of this book, the brief discussion that follows should prove helpful in understanding the magnetic properties of materials.

Scientists identify three general types of magnetism: ferromagnetism, paramagnetism, and diamagnetism. Ferromagnetic materials, like iron, nickel and certain other elements (including their compounds and alloys), have the resultant magnetic dipole moments aligned and consequently exhibit a strong magnetic field. These substances are normally referred to as magnetic materials or ferromagnetic materials. Quantum physicists suggest that a ferromagnetic substance enjoys the cooperative alignment of the electron spins of many atoms.

Materials containing rare earth elements, transition elements, or actinoid (formerly actinide) elements exhibit a property called paramagnetism. The magnetic dipole moments within such materials are randomly oriented. As a result, there is no net alignment that combines to create a significant magnetic field. In the presence of a strong external magnetic field, the individual magnetic dipole moments in a paramagnetic material tend to align and become weakly attracted by the external magnetic field. Quantum physicists suggest that this attraction is generally the result of unpaired electrons. The magnetic field of a paramagnetic substance, while observable, remains weaker than the magnetic field exhibited by ferromagnetic materials.

Finally, most common materials exhibit the property of diamagnetism. The atoms of diamagnetic substances produce only weak magnetic dipole moments. Even when placed in a strong external magnetic field, diamagnetic material exhibits only a feeble net magnetic field (if any). The diamagnetic substance is not attracted by an external magnetic field or it may be slightly repelled by this field. Quantum physicists suggest that diamagnetic substances have only paired electrons.

Today, physicists measure the strength of a magnetic field *(B)* with a unit called the *tesla* (T). One tesla is equal to one newton per ampere per meter—that is: $1\,T = 1\,N\,A^{-1}\,m^{-1}$. The tesla is a derived SI unit named in honor of the Croatian-born, Serbo-American electrical engineer and inventor Nikola Tesla (1870–1943). Scientists also use an earlier, non-SI unit for magnetic field, called the gauss (G). This unit honors the German mathematician and scientist Carl Friedrich Gauss (1777–1855). The two units of a magnetic field are related as follows: $1\,T = 10^4\,G$. The magnetic field near Earth's surface is about $10^{-4}\,T$ or $1\,G$. A powerful electromagnet produces a magnetic field of $1.5\,T$ or $1.5 \times 10^4\,G$. Finally, physicists express magnetic flux *(Φ)* in terms of webers (Wb), where $1\,Wb = 1\,T\,m^2$. The weber is a derived SI unit, honoring the German physicist Wilhelm Eduard Weber (1804–91).

DISCOVERING ELECTRICITY: FROM ANCIENT GREECE TO COLONIAL AMERICA

The study of static electricity (electrostatics) extends back to the natural philosophers of ancient Greece. About 590 B.C.E., Thales of Miletus noted that amber, when rubbed, attracted small objects (like pieces of grain or straw) to itself. The ancient Greeks called amber electron. Often used as jewelry because of its beauty, amber is typically yellow to orange in color and translucent. Amber is not a mineral, but rather fossil resin (tree sap). Despite amber's mysterious property, any comprehensive investigation of static electricity languished for centuries.

At the start of the 17th century, Gilbert's *De Magnete* not only discussed magnetism, but also stimulated renewed scientific interest in the phenomenon of electricity. Some scientists conducted experiments that generated static electricity by rubbing amber and other objects with cloth; other researchers created machines that could generate larger quantities of static electricity by friction. One such inquisitive individual was the German scientist and politician Otto von Guericke (1602–86), who made an electrostatic generator in 1660. Von Guericke's *elektrisiermachine* used a rotating ball of sulfur and a handheld piece of cloth. More adventuresome experimenters would hold a bare hand against the rotating ball of sulfur. The machine functioned reasonably well. It accumulated an electric charge, generated sparks, and even gave an overly curious researcher a mild electric shock.

The French chemist and botanist Charles du Fay (1698–1739) wrote a paper in 1733 in which he discussed the existence of two types of electricity. He called one type vitreous electricity, because it was the type of static electricity produced by rubbing glass with wool. (Vitreous is the Latin word for glass.) He called the other type resinous electricity, because it was the type of static electricity produced by rubbing amber (fossilized resin). In a variety of experiments, du Fay was able to distinguish these two types of electricity from one another because an object containing vitreous electricity would attract an object containing resinous electricity but repel another object charged with vitreous electricity. Du Fay's work gave rise to the two-fluid theory of electricity, which was quite popular in the early to mid 18th century.

Scientists regard the Leyden jar (sometimes spelled Leiden) as the technical ancestor to the modern electric capacitor. In the mid-18th century, two individuals independently developed very similar devices to store an electric charge from friction generators. The first individual was Ewald Georg von Kleist (1700–48), a German cleric and scientist. He developed his device for storing an electric charge in 1745 but failed to communicate his efforts to the scientific community in Europe. The Dutch scientist Pieter van Musschenbroek (1692–1761) independently developed a similar charge storage device at about the same time. Affiliated with the University of Leiden, van Musschenbroek published details of his invention in January 1746 and then used it to conduct numerous experiments in electrostatics. This charge-storing device soon became widely known within the scientific community as the Leyden jar.

In its most basic form, a Leyden jar consists of a glass jar that has its inner and outer surfaces coated with electrically conducting metal foil. To prevent electric arcing between the foils when charged, the conducting materials stop just short of the jar's top. A well-insulated electrode (such as a metal rod and chain arrangement) would pass through the mouth of the jar and extend down into it, enabling electrical contact with the inner foil. Scientists charge the Leyden jar by touching this central electrode to an electrostatic generator. As researchers attempted to store more and more charge in support of their investigations of electricity, many different designs appeared in the late 18th century.

The American statesman, printer, and patriot Benjamin Franklin (1706–90) was also an excellent scientist. His technical achievements were well recognized in Europe, especially in France. Franklin began his investigation of electricity in 1746. During experiments with a simple capacitor-like array of charged glass plates in 1748, he coined the term *battery*. Franklin used that term because his configuration of charge-storing devices resembled an artillery battery.

Franklin dismissed du Fay's two-fluid theory of electricity and replaced it with his own one-fluid theory. He was convinced that all the phenomena associated with the transfer of electric charge from one object to another could easily be explained by the flow of just a single electric fluid. His experiments suggested that objects either had an excess of electricity or too little electricity. To support his one-fluid theory, Franklin introduced the terms *negative* (too little) and *positive* (too much) to describe these conditions of electric charge. Franklin also coined many other electricity-related terms that are now in use. Some of these are electric shock, electrician, conductor, condenser, charge, discharge, uncharged, plus charge, and minus charge.

When asked to describe Franklin's role as a scientist, most people mention his historic kite flying experiment in which he demonstrated that lightning was electrical in nature. Although Franklin's discovery was a major accomplishment, it did not take place as generally depicted in textbooks and paintings. Flying a kite into a thunderstorm is an extremely dangerous activity. Had lightning actually struck Franklin's kite during his famous experiment on June 15, 1752, he would most likely have been electrocuted. What Franklin did during the experiment was to fly his kite into a gathering storm cloud—*before* any lightning discharges began. He detected the buildup of an electric charge within the cloud, mentally associated charge buildup with lightning, quickly ceased the experiment, and headed for safety. Franklin was definitely aware of the serious danger posed by lightning—he invented the lightning rod to protect structures.

By the end of the 18th century, scientists had determined the following important facts about electricity: (1) like electric charges repel, while unlike (positive and negative in Franklin's nomenclature) charges attract; (2) positive and negative charges appear to have equal strength; and (3) lightning is an electrical phenomenon. But no one had even a hint that the true nature of electricity involved the flow of electrons.

COUNT ALESSANDRO VOLTA

It may be difficult to imagine that all modern applications of electromagnetism descended from the availability of a dependable supply of electricity—the first of which appeared about 210 years ago. This section discusses how the Italian physicist Count Alessandro Volta (1745–1827) performed a series of key experiments in 1800, which led to the development of the first electric battery. Early batteries enabled the 19th-century research activities that led to the electricity-based Second Industrial Revolution.

This 1992 Italian stamp honors Count Alessandro Volta and illustrates the first voltaic pile (chemical battery). *(courtesy of the author)*

Volta was born on February 18, 1745, in Como, Italy. Like many other 18th-century scientists, Volta became fascinated with the subject of electricity and decided to focus his research activities on a detailed investigation of this mysterious natural phenomenon. In 1779, Volta received an appointment to become a professor of physics at the University of Pavia. He accepted this appointment and remained in this position for the next 25 years.

Volta had a friend and professional acquaintance named Luigi Galvani (1737–98), who was a physician (anatomist) in Bologna, Italy. In 1780, Galvani discovered that when a dissected frog's leg touched two dissimilar metals (such as iron and brass, or copper and zinc) at the same time, the leg twitched and contracted. Based on the observed muscular contractions, Galvani postulated that the flow of electricity in the frog's leg represented some type of animal electricity—a term he devised to suggest that electricity was the animating agent in living muscle and tissue. Galvani may have been influenced by Franklin's research, which associated lightning (a natural phenomenon) to electricity. Electricity was a frontier science in the late 18th century, so Galvani, as a scientist with a strong inclination toward anatomy, wanted to be the first investigator to successfully connect the animation of living matter with this exciting new phenomenon.

Galvani knew that Volta was also performing experiments with electricity, so he asked Volta to help validate his experiments and the hypothesis about animal electricity. This request eventually ended the amicable relationship between the two Italian scientists and made all of their subsequent interactions adversarial.

In about 1794, Volta began to explore the question of whether the electric current in the twitching frogs' legs was a phenomenon associated with biological tissue (as Galvani postulated) or actually the result of contact between two dissimilar metals. Ever the careful physicist, he tested two dissimilar metals alone without a frog's leg or other type of living tissue. He observed that an electric current appeared and continued to flow. The frog's leg had nothing to do with the current flow. Volta's conclusions dealt a mortal blow to Galvani's theory of animal electricity.

Galvani did not accept Volta's conclusions, and the two Italian scientists engaged in a bitter controversy that soon involved other famous scientists from across Europe. The French physicist Charles-Augustin de Coulomb (1736–1806) supported Volta's work and conclusions. As additional experimental evidence began to weigh heavily in Volta's favor, Galvani died a broken and bitter man on December 4, 1798. To the end, Galvani clung to his belief that electricity was inseparably linked to biology. Like many other investigators of the day, Galvani regarded electricity as a natural agent that promoted vitality.

The professional disagreement with Galvani encouraged Volta to perform additional experiments. In 1800, Volta developed the voltaic pile—the first chemical battery. From a variety of experiments, Volta determined that in order to produce a steady flow of electricity he needed to use silver and zinc as the most efficient pair of dissimilar metals. He initially made

ELECTRICITY AND LIVING THINGS

During the 1790s, the Italian physiologist Luigi Galvani actively explored a phenomenon he called animal electricity—a postulated life-giving electric force presumed capable of animating inanimate matter. Although Alessandro Volta eventually proved that animal electricity did not exist, Galvani's research did anticipate the discovery of nervous impulses (which travel through the living body) and the existence of tiny electric currents in the brain (which are now noninvasively measured in support of medical research and diagnosis).

This 1991 Italian stamp depicts the Italian physiologist Luigi Galvani and his experiments with frog legs. *(courtesy of the author)*

Electrotherapy was practiced by medical doctors and scientists from the late 1790s up through the mid-20th century. Physicians used batteries or small electric generators in their attempts to apply electric currents to cure various diseases and mental conditions. Since few reliable cures were actually achieved, electrotherapy has all but vanished from the landscape of modern medicine.

In the early 1800s, scientists and physicians were intrigued by the elusive boundary between life and death. They performed a great deal of experimental work attempting to resuscitate drowning victims with procedures that often included the use of electricity. Historians suggest that these activities and Galvani's pursuit of animal electricity may have provided the intellectual stimulus for the British author Mary Shelley's (1797–1851) iconic horror story, *Frankenstein,* which was first published in 1818.

Contemporary medical practitioners employ a variety of sensitive instruments to measure the body's own electrical impulses. With such electrical signals, physicians can routinely monitor the heart, brain, and other important organs. In addition, pacemakers help correct irregular heartbeats, and defibrillators can help restore the heart's natural rhythm after a person has experienced cardiac arrest or a dangerous arrhythmia.

individual cells by placing a strip of zinc and silver into a cup of brine. He then connected up several cells to increase the voltage. Finally, he created the first voltaic pile (chemical battery) by alternately stacking up discs of silver, zinc, and brine—soaked heavy paper—quite literally in a pile. Soon, scientists all over Europe copied and improved upon Volta's invention. The early batteries provided a steady, dependable flow of electricity (direct current) for numerous, pioneering experiments. Volta's battery sponsored a revolution in the scientific investigation of electromagnetism.

In 1810, Emperor Napoléon of France acknowledged Volta's great accomplishment and made him a count. Volta died in Como, Italy, on March 5, 1827. Modern batteries are distant technical descendants of Volta's electric pile. In their numerous shapes and forms, modern batteries continue to electrify civilization. Scientists named the SI unit of electric potential difference and electromotive forces the volt (V) to honor Volta's great contributions to science.

QUANTIFYING THE ELECTRIC NATURE OF MATTER

This section discusses the contributions of three key individuals, Coulomb, Ampère, and Ohm, who helped the scientific community better understand electricity. Their scientific contributions paved the way for the electricity-based Second Industrial Revolution.

The French military engineer and scientist Charles-Augustin de Coulomb performed basic experiments in mechanics and electrostatics in the late 18th century. In particular, he determined that the electrostatic force, which one point charge applies to another, depends directly on the amount of each charge and inversely on the square of their distance of separation. In his honor, the SI unit of electric charge is called the coulomb (C). Scientists define one coulomb as the quantity of electric charge transported in one second by a current of one ampere (A).

Building upon Coulomb's important work, experiments by physicists in the 20th century revealed that the magnitude of the elementary charge (*e*) on the proton exactly equals the magnitude of the charge on the electron. By convention, scientists say the proton carries a charge of +*e* and the electron carries a charge of −*e*. Scientists have experimentally determined the value of *e* to be 1.602×10^{-19} C.

Science historians regard André-Marie Ampère, as one of the main discoverers of electromagnetism. The gifted French scientist's defining work in this field began in 1820 and involved insightful experiments that led to the development of a physical principle called Ampere's law for static magnetic fields. His pioneering work in electromagnetism served as the foundation of the subsequent work of British experimenter Michael Faraday and American physicist Joseph Henry (1797–1878). The science of electromagnetism helped bring about the world-changing revolution in electric power applications and information technology that characterized the late 19th century. In his honor, the SI unit of electric current is called the ampere (A), or amp for short.

The German physicist George Simon Ohm (1787–1854) published the results of his experiments with electricity in 1827. His research suggested the existence of a fundamental relationship between voltage, current, and resistance. Despite the fact that his scientific colleagues dismissed these important findings, Ohm's pioneering work defined fundamental physical relationships that supported the beginning of electrical circuit analysis. Ohm proposed that the electrical resistance (R) in a material could be defined as the ratio of the voltage (V) applied across the material to the electric current (I) flowing through the material, or $R = V/I$. Today, physicists and engineers refer to this important relationship as Ohm's law. In recognition of his contribution to the scientific understanding of electricity, scientists call the SI unit of resistance the ohm (symbol Ω). One ohm of electric resistance is defined as one volt per ampere ($1\ \Omega = 1\ V/A$).

MICHAEL FARADAY'S REVOLUTION

Even though he lacked a formal education and possessed only limited mathematical skills, the British chemist and physicist Michael Faraday became one of the world's greatest experimental scientists. Of particular interest here is the fact that he made significant contributions to the field of electromagnetism. In 1831, he observed and carefully investigated the principle of electromagnetic induction—an important physical principle that governs the operation of modern electric generators and motors. In his honor, scientists named the SI unit of capacitance the farad (F). One farad is defined as the capacitance of a capacitor whose plates have a potential difference of one volt, when charged by a quantity of electricity equal to one coulomb. Since the farad is too large a unit for typical applications, submultiples—such as the microfarad (μF), the nanofarad (nF), and the picofarad (pF)—are frequently encountered in electrical engineering.

Faraday's pioneering work (ca. 1831) in electromagnetism formed the basis of modern electric power generation. He discovered that whenever there is a change in the flux through a loop of wire, an electromotive force *(emf)* is induced in the loop. Scientists now refer to this important discovery as Faraday's law of electromagnetic induction. It is the physical principle upon which the operation of an electric generator (dynamo) depends. Faraday's discovery, refined by electrical engineers into practical generators, made large quantities of electricity suddenly available for research and industrial applications. Scientists were no longer restricted to electricity supplied by chemical batteries.

Independent of Faraday, the American physicist Joseph Henry had made a similar discovery about a year earlier, but teaching duties prevented Henry from publishing his results. So credit for this discovery goes to Faraday, who not only published his results first (*Experimental Researches in Electricity, first series* [1831]) but also performed more detailed experimental investigations of the important phenomenon. However, Henry did publish a seminal paper in 1831, describing the electric motor (essentially a reverse dynamo) and its potential applications. Science historians regard the work of both Faraday and Henry during this period as the beginning of the electrified world. Clever engineers and inventors would apply Faraday's law of electromagnetic induction to create the large electric generators that supply enormous quantities of electricity. Other engineers would invent ways of using direct current (DC) and alternating current (AC) electricity to power the wide variety of practical and efficient electric motors that became the motive force of modern civilization.

Faraday was ingenious in his design and construction of experiments. However, he lacked a solid mathematics education, so translating the true significance of some of his experimental results into robust physical theory depended upon his affiliation with the Scottish theoretical physicist James Clerk Maxwell. Maxwell, a genius in his own right, competently translated the significance of Faraday's ingenious experiments into the mathematical language of physics. Their cordial working relationship resulted in a solid experimental and mathematical basis for classical electromagnetic theory.

MAXWELL'S ELECTROMAGNETIC THEORY

The Scottish mathematician James Clerk Maxwell (1831–79) is regarded as one of the world's greatest theoretical physicists. Born in Edinburgh in

1831, he was soon recognized as a child prodigy. Throughout his brilliant career, he made many contributions to science, including papers that linked the behavior of individual atoms and molecules to the bulk behavior of matter—an area of classical physics known as kinetic theory.

His seminal work, *Treatise on Electricity and Magnetism* (published in 1873), provided a detailed mathematical treatment of electromagnetism. Maxwell suggested that oscillating electric charges could generate propagating electromagnetic waves that had the speed of light as their propagation speed. Maxwell's equations provided scientists a comprehensive insight into the physical nature of electromagnetism. His work provided a detailed model of visible light as a form of electromagnetic radiation. Maxwell also suggested that other forms of electromagnetic radiation might exist beyond infrared and ultraviolet wavelengths. Scientists, like Rudolph Hertz (1857–94) and Wilhelm Conrad Roentgen (1845–1923), would soon provide experimental evidence that proved Maxwell correct.

In 1888, the German physicist Heinrich Hertz oscillated the flow of current between two metal balls separated by an air gap. He observed that each time the electric potential reached a peak in one direction (or the other), a spark would jump across the gap. Hertz applied Maxwell's electromagnetic theory to the situation and determined that the oscillating spark should generate an electromagnetic wave that traveled at the speed of light. He also used a simple loop of wire, with a small air gap at one end, to detect the presence of electromagnetic waves produced by his oscillating spark circuit. With this pioneering experiment, Hertz produced and detected radio waves for the first time. He also demonstrated that (as predicted by Maxwell) radio waves propagated at the speed of light. Hertz showed that radio waves were simply another form of electromagnetic radiation, similar to visible light and infrared radiation, save for their longer wavelengths and lower frequencies. Hertz's research formed the basis of the wireless communications industry. Scientists named the SI unit of frequency the hertz (Hz) in his honor. One hertz is equal to one cycle per second.

On November 8, 1895, the German physicist Wilhelm Conrad Roentgen discovered X-rays—another form of electromagnetic radiation. With shorter wavelengths and higher frequencies, X-rays lie beyond ultraviolet radiation in the electromagnetic spectrum.

Maxwell's brilliant theoretical interpretation of electromagnetism became the intellectual bridge between the classical physics of the 19th century and the modern physics of the 20th century. When Maxwell died

THE ELECTROMAGNETIC SPECTRUM

When sunlight passes through a prism, it throws out a rainbowlike array of colors onto a surface. Scientists call this display of colors the visible spectrum. The visible spectrum represents the narrow band of electromagnetic (EM) radiation to which the human eye is sensitive.

However, the electromagnetic spectrum consists of the entire range of electromagnetic radiation, from the shortest wavelength gamma rays to the longest wavelength radio waves and everything in between. EM radiation travels at the speed of light and represents the basic mechanism for energy transfer through the vacuum of outer space.

One of the most interesting discoveries of 20th-century physics is the dual nature of electromagnetic radiation. Under some conditions, EM radiation behaves like a wave, while under other conditions it behaves like a stream of particles called photons. The tiny amount of energy carried by a photon is called a quantum of energy.

(continues)

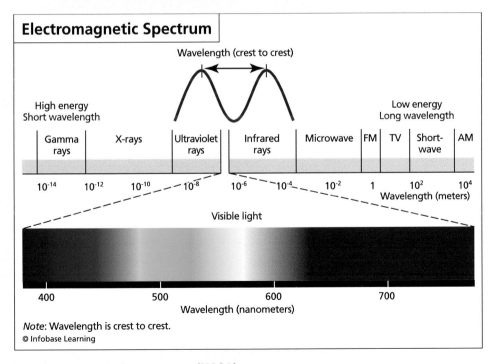

The electromagnetic spectrum *(NASA)*

(continued)

The shorter the wavelength, the more energy is carried by a particular form of EM radiation. All things in the universe emit, reflect, and absorb electromagnetic radiation in their own distinctive ways. How an object does this provides scientists a spectral signature that can be detected by remote sensing instruments. Scientists often collect spectral signatures in order to obtain information about an object's composition, density, surface temperature, and/or motion.

on November 5, 1879, in Cambridge, England, he was only 48 years old. Many scientists regard his contributions to physics comparable to those of Newton and Einstein.

ELECTRICITY AND THE SECOND INDUSTRIAL REVOLUTION

Maxwell's work revolutionized both physics and the practice of engineering. Inventors such as Thomas Edison (1847–1931) and Nikola Tesla applied electromagnetism to numerous new devices—providing people with surprising new power and comforts.

Tesla came to the United States in 1884 to accept a position with Thomas Edison's company in New York City. However, the initially cordial relationship between the two geniuses soon soured. Tesla did not respond well to Edison's authoritative management style, especially when it involved the great 19th-century controversy within the emerging electric power industry: alternating current (AC) or direct current (DC). Edison had committed himself and his company to direct current. The intellectually gifted but rebellious Tesla was the world's most talented advocate for alternating current. In a very short time, sparks began to fly as this controversy heated up and dominated electric power developments during the Second Industrial Revolution.

The big question facing the nascent electric power industry in the late 19th century was how best to transport electricity over transmission wires without incurring too great an energy (heating) loss. Tesla recognized that transporting electricity at high voltage using transformers at both the generating station (to raise the voltage) and then at the consumer end (to lower the voltage) solved the problem. However, transformers only work with AC systems. In addition to the growing professional disagreement over the choice of DC versus AC, Tesla was also embittered because Edison apparently failed to make good on a promised bonus payment for the special work Tesla performed for Edison to improve the efficiency of Edison's DC generators.

So after just a year Tesla quit his job with Edison's company and set out to prove AC was best. His departure marked the start of a bitter feud between the two geniuses, a feud that raged for decades.

Responding to his disappointing relationship with Edison, Tesla formed his own company in 1886 and called it the Tesla Electric Light and Manufacturing. His goal was to develop and market an AC motor. Unfortunately, this goal caused a major disagreement between Tesla and his financial backers, who soon relieved him of his duties at the fledgling company that bore his name. Undeterred, Tesla worked as a common laborer for the next year to support himself and to save enough money for the construction of the electromagnetic induction motor, for which he eventually received a U.S. patent. Tesla's revolutionary motor used a rotating magnetic field, rather than mechanical switches (that is, commutators) to spin the armature or rotor. A commutator is found in certain types of electric motors and generators. It is a rotary electric switch that reverses the direction of current flow in a periodic fashion. Tesla's device opened the way for the modern three-phase AC power system (generator, transformers, and motors) and for the common electrical devices found in most factories, offices, and homes.

In 1886, the American engineer and industrialist George Westinghouse (1846–1914) founded Westinghouse Electric in Pittsburgh and entered head-to-head commercial competition with Edison, who was convinced the DC electric power system he had installed in New York City could not be outdone. At that time, the effective range of delivery of DC electricity was only three miles (five km) at best. Westinghouse believed in the viability of AC and joined with Tesla to create the generators, transformers, and motors necessary to deliver AC electricity to a much larger number of customers over greater distances. In 1888, Tesla received U.S. patents for his three-phase (polyphase) system of AC generators, transformers, and motors. That same year, Tesla sold his patents for the AC motor and dynamo (generator) to Westinghouse, who hired Tesla and funded his research related to commercializing AC. Their pioneering efforts would make AC the standard for commercial electric power generation and transmission.

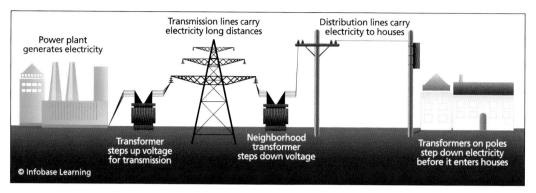

This diagram depicts the overall generation, transmission, and distribution of electricity. Alternating current (AC) electric power is generated at power plants (operated by fossil fuel, nuclear, or renewable energy sources) then moved to substations by means of high-voltage transmission lines. The grid is the network of nearly 160,000 miles (257,600 km) of high-voltage transmission lines that now crisscrosses the United States. A local distribution system (above- or belowground) of smaller, lower-voltage distribution lines carries electricity from substations and transformers to customers. *(adapted from DOE/ EIA/National Energy Education Development)*

After a number of impressive public demonstrations at the end of the 19th century, the AC system advocated by Tesla and Westinghouse soundly defeated Edison's DC system. For example, with Westinghouse's support, Tesla's AC equipment was used to illuminate the 1893 World Columbian Exhibition in Chicago. In 1895, Westinghouse won a coveted contract to use Niagara Falls to generate electricity and to deliver the generated (AC) electricity to the city of Buffalo, New York, a city about 22 miles (35 km) away. In a certain sense, these demonstrations and the victory of the Tesla-Westinghouse AC electric power system over Edison's DC electric power system serve as the pinnacle of Tesla's professional life. He gained worldwide attention, some short-term wealth, and expanded his social circle to include notables such as Mark Twain.

In 1899, Tesla moved to Colorado Springs, where he constructed a large laboratory so he could investigate lightning and conduct high-frequency, high-voltage experiments exploring the possibilities of wireless telegraphy, telephony, and even the wireless transmission of electric power. Tesla left Colorado Springs in early January 1900 and began planning his next project, called the Wardenclyffe Tower facility in Shoreham, Long Island, New York.

In early summer 1902, Tesla moved his laboratory operations from Houston Street in New York City to the Wardenclyffe facility. The site's huge 187-foot (57-m) antenna was to serve as the first station in Tesla's envisioned worldwide wireless telecommunications system. Although the structure for the radio tower was completed in 1904, Tesla never completed his planned transceiver because his financial supporters pulled out. Faced with rising debts, Tesla abandoned the project. The tower was demolished and sold for scrap in 1917. The entire experience left Tesla in permanent financial distress and a state of deep depression from which he would never recover.

Soon after the German physicist Heinrich Hertz produced and detected radio waves for the first time in 1888, Tesla, the Italian physicist Guglielmo Marconi (1874–1937), and other late 19th-century researchers began exploring the possibility of wireless communications. Tesla's radio wave research put him on a direct collision course with Marconi. In fact, Tesla always disputed the claim that Marconi invented radio. Up to World War I, Tesla engaged in an expensive (but unsuccessful) legal battle against Marconi. By 1916, Tesla was forced to file for bankruptcy and spent the rest of his life in poverty. Tesla died in New York City on January 7, 1943. Ironically, in the year of Tesla's death (1943), the U.S. Supreme Court ruled that Tesla's patents for the radio superseded those of Marconi.

Tesla was clearly one of the greatest engineers of all time. His inventions helped to electrify the modern world, yet his genius, with its obsessive-compulsive dark side, brought him neither wealth nor contentment. In the mid-1950s, the international scientific community named the SI unit of magnetic flux density the tesla (T) in his honor. One tesla is equal to one weber per square meter.

Thanks to geniuses like Tesla and Edison, the use of electricity pervades modern life. The digital (or information) revolution is a generic expression that actually encompasses several major technology shifts that occurred in the mid to late 20th century. The first part of this process was the discovery of the transistor in the late 1940s; the second part involved the application of the transistor in the subsequent microelectronics revolution. The integrated circuit was nothing short of a technical miracle in materials science and engineering. This device accelerated the exponential development and application of digital computers and microprocessors.

The discoveries about the nature of electromagnetism during the 19th century reached an exciting scientific climax in 1897. In that year, the British physicist Sir Joseph John (J. J.) Thomson (1856–1940) published the results of his experiments that clearly demonstrated the existence of the electron—the first known subatomic particle.

(continues on page 140)

ELECTRIC VEHICLES

The electric vehicle (EV) is not a recent development. It has been around for well over 100 years. In fact, at the beginning of the 20th century the number of electric vehicles and steam-powered vehicles in the United States far exceeded the number of horseless carriages powered by internal combustion engines.

The story of the electric vehicle begins in Europe in the 1830s. Creative individuals in Holland, France, and Great Britain started using electric batteries and motors to propel horseless carriages. As the capacity and lifetime of batteries improved, so did the performance of these early electric vehicles, which were mostly nothing more than electrified horseless carriages and surreys.

It was not until 1895 that American inventors gave EVs some serious attention. Many innovations soon followed, and the EV population quickly increased in the late 1890s and early 1900s. In 1897, the first commercial EV application

An engineer takes the all-electric Tesla roadster for a test-drive at the Argonne National Laboratory in 2010. Powered by lithium-ion batteries, the vehicle's 375-volt AC induction air-cooled electric motor delivers 288 horsepower (215 kW) at 5,000 to 6,000 RPM. The Tesla roadster can accelerate from 0 to 60 MPH (96.6 km/h) in 3.9 seconds; has a maximum speed of 125 MPH (201 km/h); and a range of about 245 miles (395 km) on a battery charge. *(DOE/ANL)*

occurred in New York City and involved a fleet of taxicabs. With paved streets and short distances between destinations, the urban environment proved ideal for electric vehicles.

By the turn of the century, motor vehicles, now available in steam, electric, or gasoline versions, became popular in the United States—especially for wealthy individuals. From 1899 to 1912, interest in EVs peaked, and they outsold all other types of cars. Companies like the Baker Motor Vehicle Company of Cleveland, Ohio, and Detroit Electric produced a variety of interesting electric vehicles. One Baker vehicle, a racer called the Torpedo, set a speed record of 75 MPH (121 km/h). One EV owner was the American inventor Thomas Edison who subsequently became involved in manufacturing improved nickel-iron batteries. During this period, a Detroit Electric test EV operated for 211 miles (340 km) on a single charge.

Electric vehicles had many advantages over their competitors in the early 1900s. They did not have the vibration, smell, and noise associated with internal combustion engine automobiles. The electric vehicle was the preferred choice of many because it did not require a significant manual effort to start, as occurred with the hand crank on early gasoline-powered vehicles. Changing gears on these gasoline-powered cars typically proved to be the most difficult part of driving. In contrast, electric vehicles and steam-powered vehicles did not require such gear changes. Steam-powered cars had less range than a typical EV of the period. Drivers of steam-powered cars usually needed to stop frequently and replace the supply of water for the boiler. Steam-powered cars also suffered from long start-up times of up to 45 minutes on cold mornings.

Most early electric vehicles were ornate massive carriages designed for wealthy patrons. They had plush interiors, manufactured with expensive materials, and averaged $3,000 by 1910. Production peaked in 1912, and the EV enjoyed success into the 1920s. But technical developments caused the decline of electric vehicle use. By the 1920s, the United States had developed a better system of roads that connected cities. Longer-range vehicles were needed to travel between cities. The discovery of rich oil fields in Texas significantly reduced the price of gasoline. This fuel soon became affordable for use by the average American consumer. In 1912, the creative American inventor Charles Franklin Kettering (1876–1958) developed the electric starter for gasoline-powered engines, eliminating the need for the troublesome hand crank. Finally,

(continues)

(continued)

the American industrialist Henry Ford introduced the movable assembly line. His innovative approach to the mass production of gasoline engine–powered automobiles made them widely available and affordable. By contrast, the price of the less efficiently produced electric vehicles continued to rise. In 1912, a typical electric roadster sold for about $1,750, while a gasoline-powered automobile sold for just $650.

Similar price differentials exist in today's automobile marketplace. Most people favor the use of environmentally friendly vehicles. But dire economic circumstances can force consumers to make less expensive choices. Although plug-in hybrid electric vehicles (PHEVs) are now in a pre-commercial stage of development, they represent an interesting personal transportation option for the future. PHEVs can be charged with electricity like pure all-electric vehicles and run under engine power like hybrid vehicles. The combination of options offers increased driving range with potentially large fuel savings and emission reductions. Like hybrid vehicles, PHEVs are powered by two basic energy sources: an energy conversion unit (such as an internal combustion engine or a fuel cell) and an energy storage device (most probably electric storage batteries). Engineers suggest that the PHEV's energy conversion unit may be powered by gasoline, diesel, compressed natural gas, hydrogen, or other fuels. The energy storage unit (rechargeable batteries) may be charged by plugging into a standard 110-volt electric outlet, as well as being charged by the energy conversion unit.

(continued from page 137)

THE CONDUCTION OF ELECTRICITY

Scientists now recognize that some substances (such as silver and copper) readily conduct the flow of electricity and call these materials electrical conductors. Other materials (such as wood and most plastics) resist the flow of electrons and are called electrical insulators. The basic difference between conductors and insulators lies in their atomic structures. A material that is a good electrical conductor has one or several outer (valence) electrons that are loosely attached to the parent atomic nucleus and, therefore, available to freely wander through the material under the influence of an applied voltage. The situation is quite different for a material that is a good electrical insulator. In this case, almost every electron remains tightly bound to its parent atom. So, even when a voltage difference is applied across the material, the outer electrons cannot wander freely through it.

When scientists discuss the property of electrical conductivity, they often include a third common type of material called a semiconductor. A semiconductor is a solid crystalline material (such as silicon [Si] or germanium [Ge]) that has a typical electrical conductivity intermediate between the values of good electrical conductors and insulators. Semiconductors can carry electric charges but not very well. The conductance of silicon is about 1 million times less than the electrical conductance of copper.

With an electrical conductivity (symbol σ) of 63×10^6 siemens per meter (S/m) at a temperature of 540 R (300 K), silver is an excellent conductor of electricity. Copper follows closely behind with an electrical conductivity of about 60×10^6 S/m at 540 R (300 K). For economic and resource availability reasons, copper is widely used in the distribution of electricity. Gold has an electrical conductivity of 45×10^6 S/m at 540 R (300 K) and aluminum about 38×10^6 S/m at 540 R (300 K). Please note that the electrical conductivity of metals is temperature dependent and decreases significantly with increasing temperature.

The electrical resistivity (symbol ρ) of a material is the reciprocal of that material's electrical conductivity. Silver has an electrical resistivity of 1.6×10^{-8} ohm-meter (Ω-m) at a temperature of 540 R (300 K). At 540 R (300 K), pure silicon has an electrical resistivity value of 2500 Ω-m; a typical n-type silicon semiconductor material has an electrical resistivity of 8.7×10^{-4} Ω-m; and a typical p-type silicon semiconductor material has an electrical resistivity of 2.8×10^{-3} Ω-m. Engineers call silicon that has been doped with (electron) donor atoms (such as phosphorus) n-type semiconductors; silicon that has been doped with (electron) acceptor atoms (like aluminum) p-type semiconductors. Finally, they regard glass and fused quartz as good electrical insulators. The electrical resistivity of glass at 540 R (300 K) ranges from 10^{10} to 10^{14} Ω-m. Fused quartz has an approximate electrical resistivity of 1×10^{16} Ω-m.

SIR JOSEPH JOHN (J. J.) THOMSON DISCOVERS THE ELECTRON

At the end of the 19th century, the British physicist Sir Joseph John (J. J.) Thomson made a monumental discovery. Using Crookes tubes (cold cathode tubes), he performed experiments that demonstrated the existence of the first known subatomic particle, the electron.

Thomson's discovery implied that the previously postulated indivisible, a solid atom, was really divisible and contained smaller parts. When he announced his finding during a lecture at the Royal Institution in 1897, his audience was initially reluctant to accept the idea that objects smaller than an atom could exist. But Thomson was correct, and he received the 1906 Nobel Prize in physics for his landmark discovery. He was knighted in 1908.

Thomson originally referred to the tiny subatomic particles he discovered as corpuscles. Several years earlier (in 1891), the Irish physicist George Johnstone Stoney (1826–1911) suggested the name electron for the elementary charge of electricity. In time, Stoney's term became the universal name by which Thomson's tiny negatively charged corpuscles became known. The existence of the electron showed that (some) subatomic particles carried a fundamental (quantized) amount of electric charge.

In 1898, Thomson became the first scientist to put forward a technical concept concerning the interior structure of the atom. He suggested that the atom was actually a distributed positively charged mass with an appropriate number of tiny electrons embedded in it, much like raisins in a plum pudding. Because there were still so many unanswered questions, not many physicists rushed to embrace Thomson's model of the atom—commonly referred to as the plum pudding model. But the Thomson atom, for all its limitations, started the New Zealand–born British physicist Baron Ernest Rutherford (1871–1937) and other atomic scientists thinking about the structure within the atom.

Today, scientists regard the electron (e) as a stable elementary particle with a fundamental unit of negative charge (-1.602×10^{-19} C) and a rest mass of just 2.009×10^{-30} lbm (9.109×10^{-31} kg)—a mass approximately 1/1,837 that of a proton. Scientists now recognize that electrons surround the positively charged nucleus of an atom and determine a substance's chemical properties and behavior. The manipulation and flow of electrons support today's information-dependent global civilization.

CHAPTER 8

Discovering the Equivalence of Energy and Matter

This chapter presents some of the fundamental concepts and physical laws that form the basis of nuclear energy applications. Emphasis is placed on how a nuclear reactor operates and the use of nuclear reactors for the generation of electric power. According to the *2010 Electric Power Annual* published by the U.S. Energy Information Administration (EIA), 20 percent of the nation's 4.1 trillion kilowatt-hours of electricity came from nuclear energy. The chapter concludes with a brief discussion about contemporary efforts to achieved controlled nuclear fusion.

THE PILLARS OF TWENTIETH-CENTURY PHYSICS

The first four decades of the 20th century were a golden age of intellectual achievement in science. The German physicist Max Planck (1858–1947) introduced quantum theory in 1900 and the German-Swiss-American physicist Albert Einstein (1879–1955) presented special relativity in 1905 and general relativity in 1915. Their pioneering efforts constitute the two foundational pillars of 20th-century physics and profoundly influenced many aspects of today's technology-dependent global civilization.

Physicists often refer to classical physics as the body of scientific knowledge bracketed by the mechanics of Sir Isaac Newton and the electromagnetic theory of James Clerk Maxwell. They regard quantum theory and relativity as the start of the era of modern physics. Although now

The Tennessee Valley Authority's (TVA) Sequoyah Nuclear Plant is located beside the Chickamauga Reservoir in Soddy-Daisy, Tennessee. Each of the power plant's two pressurized water reactor units is capable of generating more than 1,160 megawatts of electricity. Tall, hyperbolic cooling towers reject waste heat from the plant's condensers in a safe and environmentally benign manner. Unlike fossil-fuel generating plants, nuclear plants do not release carbon dioxide to the atmosphere. *(TVA)*

about one century old, the term *modern physics* is still an appropriate one, since a great deal of 21st-century physics owes its origins to the foundational work of Planck and Einstein. However, recalling the enigmas of dark matter and dark energy discussed earlier in this book, scientists anticipate many dramatic changes in the definition of what constitutes modern physics, as they continue to unravel the mysteries of matter and energy during the remainder of this century.

Planck's pioneering work supported a more comprehensive understanding of matter and energy at the smallest imaginable scale. Specifically, Planck suggested that matter radiates energy not continuously, but rather in discrete chunks or quanta. When Einstein developed his theory of the photoelectric effect in 1905, he embraced Planck's notion of discrete energy packets by postulating that light itself is composed of energetic particles called photons. Einstein's special relativity included the revolutionary notion of the equivalence of mass and energy. His deceptively simple mass-energy formulation gave scientists the means of understanding the complex processes taking place in the atomic nucleus during fission, fusion, or radioactive decay. Energy is liberated in such nuclear reactions because equivalent amounts of matter disappeared. Finally, Einstein's general relativity provided scientists a brilliant new way of understanding matter's gravitational influence on cosmic scales.

In formulating special relativity, Einstein proposed two fundamental postulates:

- First postulate of special relativity: The speed of light (*c*) has the same value for all (inertial-reference-frame) observers, independent and regardless of the motion of the light source or the observers.
- Second postulate of special relativity: All physical laws are the same for all observers moving at constant velocity with respect to each other.

From the theory of special relativity, Einstein concluded that only a zero-rest-mass particle, like a photon, could travel at the speed of light. One major consequence of special relativity is the equivalence of mass and energy. This important relationship is succinctly expressed in Einstein's famous formula: $E = m c^2$, where E is the energy equivalent of an amount of matter (m) that is annihilated or converted completely into pure energy and *c* is the speed of light. Among many other important physical insights, this equation was the key scientists needed to understand energy release in such nuclear reactions as fission, fusion, and radioactive decay. Einstein's special relativity theory became one of the foundations of physics in the 20th century.

Einstein introduced his general theory of relativity in 1915. He used this development to describe the spacetime relationships of special relativity for cases where there was a strong gravitational influence such as white dwarf stars, neutron stars, and black holes. One of Einstein's

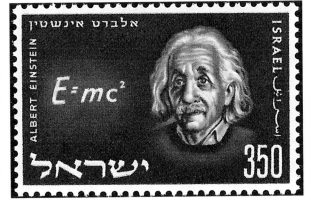

Appearing with Einstein on this 1956 stamp from Israel is his famous mass-energy equivalence equation—arguably the most well-known equation in all physics. *(courtesy of the author)*

PHOTOELECTRIC EFFECT

When sufficiently energetic photons fall upon the surface of certain substances, especially shiny metals, orbiting electrons can completely absorb the incident photons and then depart their parent atoms in the target material. The phenomenon is called the photoelectric effect, a phenomenon that plays an important role in many modern electronic devices.

While experimenting with ultraviolet radiation in 1887, the German physicist Heinrich Hertz observed that the incident ultraviolet light was releasing electric charges (later called electrons) from the surface of a thin metal target. Unfortunately, Hertz did not recognize the significance of this phenomenon nor did he pursue further investigation of the photoelectric effect.

In 1905, Albert Einstein wrote an important paper about the photoelectric effect, linking the phenomenon with Max Planck's idea of quantum packets of energy called photons. Einstein received the 1921 Nobel Prize in physics for his explanation of the photoelectric effect. His hypothesis involved the conservation of energy principle and the postulation of single photon absorption by atoms in the target material. By successfully explaining this important phenomenon, Einstein helped establish the field of quantum mechanics.

conclusions was that gravitation is not really a force between two masses (as Newtonian mechanics suggests) but rather arose as a consequence of the curvature of space and time. In a four-dimensional universe (described by three spatial dimensions [x, y, and z] and time [t]), spacetime becomes curved in the presence of matter, especially large concentrations of matter. The fundamental postulate of general relativity states that the physical behavior inside a system in free fall is indistinguishable from the physical behavior inside a system far removed from any gravitating matter (that is, the complete absence of a gravitational field). This very important postulate is called Einstein's principle of equivalence.

RADIOACTIVE DECAY

For many atoms, the protons and neutrons arrange themselves in such a way that their nuclei become unstable and spontaneously disintegrate at different, but statistically predictable, rates. As an unstable nucleus attempts to reach stability, it emits various types of nuclear radiation.

Early nuclear scientists, such as the Polish-French radiochemist Marie Curie (1867–1934), called the first three forms of nuclear radiation alpha rays, beta rays, and gamma rays. They assigned the letters of the Greek alphabet to these phenomena in the order of their discovery. At the time, the precise nature of the interesting rays that emanated from naturally radioactive substances, such as radium and polonium, was not clearly understood.

Scientists use the rather general term *radiation* to describe the propagation of waves and particles through the vacuum of space. The term includes both electromagnetic radiation and nuclear particle radiation. As discussed in chapter 7, electromagnetic radiation has a broad continuous spectrum that embraces radio waves, microwaves, infrared radiation, visible radiation (light), ultraviolet radiation, X-rays, and gamma rays. Photons of electromagnetic radiation travel at the speed of light. With the shortest wavelength (λ) and highest frequency (ν), the gamma ray photon is the most energetic. By way of comparison, radio wave photons have energies between 10^{-10} and 10^{-3} electron volt (eV), visible light photons between 1.5 and 3.0 eV, and gamma ray photons between approximately 1 and 10 *million* electron volts (MeV).

One of the most important defining characteristics of radiation is energy. So, scientists need a convenient measure of energy to make comparisons easier. For ionizing radiation, a common unit is the electron volt (eV)—the kinetic energy that a free electron acquires when it accelerates across an electric potential difference of one volt. The passage of ionizing radiation through matter causes the ejection of outer (bound) electrons from their parent atoms. An ejected (or "free") electron speeds off with its negative charge and leaves behind the parent atom with a positive electric charge. Scientists call the two charged entities an ion pair. On the average, it takes about 25 electron volts (25 eV) to produce an ion pair in water. Ionizing radiation can damage or harm many material substances, including living tissue, by rapidly creating a large number of ion pairs in a small volume of matter.

The term *nuclear radiation* refers to the particles and electromagnetic radiations emitted from atomic nuclei as a result of nuclear reaction processes, including radioactive decay and fission. The most common forms of nuclear radiation encountered in nuclear technology applications are alpha particles, beta particles, gamma rays, and neutrons. However, nuclear radiation may also appear in other forms, such as energetic protons from accelerator-induced reactions or the spontaneous fission of heavy nuclei. When discussing the biological effects of nuclear radiation, scientists include X-rays in the list of ionizing radiations. Although energetic photons, X-rays are not (as often misunderstood) a form of nuclear radiation, since they do not originate from processes within the nucleus of an atom.

Radioactivity is the spontaneous decay or disintegration of an unstable nucleus. The emission of nuclear radiation, such as alpha particles, beta particles, gamma rays, or neutrons, usually accompanies the decay process. There are two commonly encountered units of radioactivity: the curie (Ci) and the becquerel (Bq). The curie is the traditional unit used by scientists to describe the intensity of radioactivity in a sample of material. One curie (Ci) of radioactivity is equal to the disintegration or transformation of 37 billion (37×10^9) nuclei per second. The unit honors the French scientists Marie and Pierre (1859–1906) Curie who discovered radium (Ra) in 1898. The curie corresponds to the approximate radioactivity level of one gram of pure radium, a convenient, naturally occurring radioactive standard that arose in the early days of nuclear science. The becquerel is the SI unit of radioactivity. One becquerel (Bq) corresponds to the disintegration (or spontaneous nuclear transformation) of one atom per second. This unit honors the French physicist Antoine-Henri Becquerel (1852–1908), who discovered radioactivity in 1896.

When the nuclei of some atoms experience radioactive decay, a single nuclear particle or gamma ray appears. However, when other radioactive nuclides decay, a nuclear particle and one or several gamma rays may appear simultaneously. Since the curie and the becquerel both describe radioactivity in terms of the number of atomic nuclei that transform or disintegrate per unit time, they do not necessarily correspond to the actual number of nuclear particles or photons emitted per unit time. To obtain the latter, scientists usually need a little more information about the particular radioactive decay process of interest. For example, each time an atom of the radioisotope cobalt-60 decays the nucleus emits an energetic beta particle (of approximately 0.31 MeV energy) along with two gamma ray photons: one at 1.17 MeV energy of 1.17 MeV and the other at 1.33 MeV. During the decay process, the radioactive cobalt-60 atom ($^{60}_{27}$Co) transforms into the stable nickel-60 atom ($^{60}_{28}$Ni).

Scientists cannot determine the exact time that a particular nucleus within a collection of the same type of radioactive atoms will decay. However, as first quantified by Ernest Rutherford and the British chemist Frederick Soddy (1877–1965), the average behavior of the nuclei in a very

RADIOISOTOPES FOR DEEP-SPACE EXPLORATION

Reliable, abundant, and portable energy has been and remains a key factor in the successful exploration of the distant regions of the solar system by NASA spacecraft. Space nuclear power systems, using the radioisotope plutonium-238, offer several distinct advantages: compact size, long operating lifetimes, the ability to function in hostile environments, and the ability to operate independent of the spacecraft's distance from the Sun or its the orientation to the Sun. Autonomy from the Sun represents a unique technical advantage of space nuclear power systems. At present, nuclear energy is the only practical way to provide electric power to a spacecraft that must function in deep space while successfully conducting scientific missions that extend for years.

Within the U.S. national space program, the use of space nuclear power technology has and continues to emphasize the safe and responsible use of the long-lived radioisotope, plutonium-238, as a heat source. The latest example involves NASA's *New Horizons* mission to Pluto and beyond. Launched in January 2006, this nuclear-powered spacecraft will encounter Pluto in the summer of 2015 and then continue on to investigate one or more Kuiper Belt objects (KBOs) from 2016 to about 2020. The Kuiper Belt is a vast reservoir of icy objects that are located just outside of Neptune's orbit and extend about 50 astronomical units (AUs) from the Sun. Most astronomers now regard Pluto and its large moon Charon as among the largest ice dwarfs in the Kuiper Belt.

Electrical power for the *New Horizons* spacecraft and its science instruments is provided by a single, plutonium-238 fueled radioisotope thermoelectric generator (RTG), supplied for the mission by the Department of Energy (DOE). The RTG is an inherently simple and reliable device for producing spacecraft electric power. The basic RTG consists of two fundamental components: a radioisotope heat source (fuel and containment) and the thermoelectric generator, an energy conversion device that directly transforms heat into electricity. In this case, heat-to-energy conversion takes place by means of the Seebeck thermoelectric effect. The Russian-German scientist Thomas Johann Seebeck (1770–1831) discovered the phenomenon in 1821. He observed that if two dissimilar metals were joined at two locations, which were maintained at different temperatures, an electric current would flow in a loop. This physical principle provides aerospace engineers a convenient and dependable way to produce a flow of electric

(continues)

(continued)

current without resorting to a more complex electromagnetic generator system that has moving parts.

Long-duration, deep-space missions require a radioisotope fuel with a sufficiently long half-life, a reasonable energy output per disintegration, and minimal radiation shielding needs. With a half-life of 87.7 years, the alpha-emitting radioisotope plutonium-238 is a good technical choice.

This artist's rendering shows NASA's *New Horizon* spacecraft, as it approaches the dwarf planet Pluto and its three moons in the summer of 2015. As the spacecraft flies past, its suite of scientific instruments will carefully study Pluto and the large moon Charon. The 7-foot (2.1-m) dish antenna will allow the spacecraft to communicate back to Earth across 4.7 billion miles (7.5 billion km) of space. A radioisotope thermoelectric generator (gray-colored finned-cylinder on right side of spacecraft) provides electric power. *(NASA/JHUAPL/SwRI)*

large sample of a particular radioisotope can be predicted quite accurately using statistical methods. Their work became known as the law of radioactive decay. The radiological half-life ($T_{1/2}$) is the period of time required for one-half the atoms of a particular radioactive isotope to disintegrate or decay into another nuclear form. The half-life is a characteristic property of each type of radioactive nuclide. Experimentally measured half-lives vary from millionths of a second to billions of years.

A decay chain is the series of nuclear transitions or disintegrations that certain families of radioisotopes undergo before reaching a stable end nuclide. Following the tradition introduced by the early nuclear researchers, scientists call the lead radioisotope in a decay chain the parent nuclide, and subsequent radioactive isotopes in the same chain, the daughter nuclide, granddaughter nuclide, great-granddaughter, and so forth until the last, stable isotope for that chain occurs. Some decay chains are very simple; others are quite complex.

The naturally occurring radioisotope uranium-235 has a half-life of 7.04×10^8 years; decays into the radioisotope thorium-231; and emits a 4.39 MeV alpha particle as its major nuclear radiation during the decay process. The naturally occurring radioisotope uranium-238 has a half-life of 4.47×10^9 years; decays into the radioisotope uranium-234; and emits a 4.20 MeV alpha particle as its major radiation during the decay process.

The human-manufactured radioisotope plutonium-238 has a half-life of 87.7 years; decays into the radioisotope uranium-234; and emits a 5.49 MeV alpha particle as its major radiation during the decay process. The radioactivity level of one gram of pure plutonium-238 is approximately 17.2 Ci (6.36×10^{11} Bq). The human-manufactured radioisotope plutonium-239 has a half-life of 24,100 years; decays into the radioisotope uranium-235; and emits a 5.15 MeV alpha particle as its major radiation during the decay process. The radioactivity level of one gram of pure plutonium-239 is about 61.3 mCi (2.27×10^9 Bq).

NUCLEAR FISSION

During the process of nuclear fission, the nucleus of a heavy element, such as uranium (U) or plutonium (Pu), is bombarded by a neutron, which it then absorbs. The resulting compound nucleus is unstable and soon breaks apart (or fissions), forming two lighter nuclei (called fission products) and releasing additional neutrons.

In a properly designed nuclear reactor, the fission neutrons are used to sustain the fission process in a controlled chain reaction. The fission process is also accompanied by the release of a large amount of energy, typically 200 million electron volts (MeV) per reaction. Much of this energy appears as the kinetic energy of the fission product nuclei, which kinetic energy is then converted to heat as the fission products slow down in the reactor fuel material. Engineers use a circulating coolant to remove heat from the reactor's core and then apply the heated coolant in the generation of electricity or in industrial processes requiring heat.

Energy is released during the nuclear fission process because the total mass of the fission products and neutrons after the reaction is less than the total mass of the original neutron and heavy fissionable nucleus that absorbed it. Einstein's mass-energy equivalence formula determines the specific amount of energy released in a fission reaction. Nuclear fission can occur spontaneously in heavy elements but is usually caused when nuclei absorb neutrons. In some (rarer) circumstances nuclear fission may also be induced by very energetic gamma rays during a process called photofission.

The most important fissionable (or fissile) materials are the radioisotopes uranium-235 and plutonium-239. Uranium-235 occurs as a very small percent (about seven atoms in 1,000) of all uranium atoms found in nature. The bulk of the natural uranium atoms are the radioisotope uranium-238, which is not fissionable under most conditions found in commercial nuclear reactors. Scientists call uranium-238 a fertile nuclide, meaning that when it absorbs a neutron, instead of undergoing fission, it eventually transforms (by way of the formation and decay of neptunium-239) into another fissile nuclide, plutonium-239. Metric tons of plutonium-239 now exist because of the neutron capture and elemental transformation process that takes place in the core of nuclear reactors.

In concept, a nuclear power plant is quite similar to a fossil fuel (coal, petroleum, or natural gas) power plant, since both use heat engine technology to generate electricity. However, there is one fundamental difference, the source of heat. The reactor in a nuclear power plant performs the same function as the combustion of fossil fuel in the other types of thermal electric power plants. Through the controlled splitting of fissile nuclei (such as uranium-235) in its core, a nuclear reactor releases an enormous amount of heat in a relatively small volume. Depending on the specific design of the nuclear power plant, the heat liberated by nuclear fission reactions eventually transforms liquid water into the high temperature steam that spins turbine-generators.

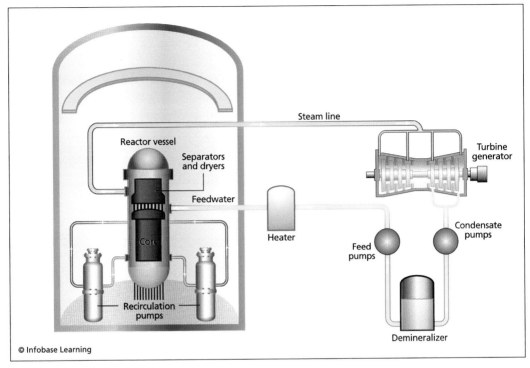

Steam line

Reactor vessel

Separators
and dryers

Turbine
generator

Feedwater

Core

Heater

Condensate
pumps

Feed
pumps

Recirculation
pumps

Demineralizer

© Infobase Learning

The components of a typical commercial boiling water reactor *(NRC)*

Nuclear reactors are basically machines that contain and control fission chain reactions, while releasing heat at a controlled rate. In most contemporary nuclear electric power plants, the reactor serves as the heat source that turns water into the steam used in a thermodynamic (heat engine) cycle, which results in the spinning of a turbine-generator system. Most of the commercial nuclear power reactors in the world today are either the boiling water reactor (BWR) or the pressurized water reactor (PWR). Nuclear engineers in the United States, Japan, France, and elsewhere are designing advanced versions of these systems for continued nuclear power generation activities throughout this century.

The figure above illustrates the basic components in a typical commercial boiling water reactor nuclear power plant. The reactor's core releases heat through controlled nuclear fission reactions. This heat is absorbed by the reactor coolant (very pure water) as it flows upward through the reactor core. A mixture of steam and water leaves the top of the reactor core and enters the two stages of moisture separation where water droplets are removed before the steam is allowed to enter the steam line. The

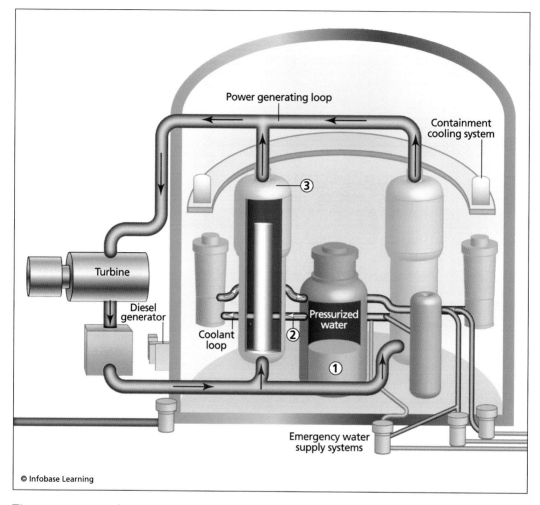

The components of a typical commercial pressurized light-water reactor *(NRC)*

steam line transports the energy-rich steam to the main turbine, where it expands and spins the turbine-generator system, producing electricity. The energy-depleted steam is now exhausted to the condenser, where it releases any residual heat and experiences a change of state back to water. After demineralization and preheating activities, the resultant water (called feedwater) is pumped back to the reactor vessel. The reactor's core contains nuclear fuel assemblies that are cooled by water, which is force-circulated by electrically powered pumps. The system has many safety features including emergency cooling systems operated by independent electric power supplies (typically on-site diesel generators).

The figure on page 154 shows the basic components in a typical pressurized water reactor. In this nuclear power system, the reactor core (denoted as 1) supplies heat. Pressurized light-water coolant (light water means ordinary H_2O as opposed to heavy water D_2O used in other types of reactor systems) circulates in the primary coolant loop (denoted as 2), taking heat away from the core and making it available to the steam generator (denoted as 3). Inside the steam generator, heat from the primary coolant loop vaporizes the water in a secondary coolant (power-generating) loop producing steam. The energy-rich steam in the power-generating loop travels to the turbine and expands, causing the turbine-generator system to spin and produce electricity. The energy-depleted steam in the secondary coolant loop now exhausts to the condenser, where it releases any residual heat and transforms from a vapor back to a liquid. The resulting water is removed from the condenser by a series of pumps. After preheating, the liquid water travels back to the steam generator to renew its journey as steam through the power-generating loop. Electrically powered pumps circulate the pressurized water in the primary coolant loop through the reactor's core. The system has many safety features, including emergency cooling systems operated by independent electric power supplies (typically on-site diesel generators).

Nuclear engineers recognize that each type of light-water reactor has certain technical advantages and disadvantages. That is why both the BWR and the PWR provide commercial nuclear power–generating service around the world. The actual layout of any of these nuclear power plants is very complex. Each plant contains a great deal of sophisticated equipment—the detailed discussion of which goes far beyond the scope of this chapter.

NUCLEAR FUEL CYCLE

The nuclear fuel cycle for typical light-water reactors consists of front end steps that lead to the preparation of uranium for use as fuel for reactor operation and back end steps that are necessary to safely manage, prepare, and dispose of the highly radioactive spent nuclear fuel. Chemical processing of the spent fuel material takes place to recover the remaining fractions of fissile nuclides, namely the uranium-235 from the original fuel charge that did not undergo fission and any plutonium-239 produced in the reactor as a result of the non-fission capture of neutrons by uranium-238. It is technically feasible to recover both the uranium-235 and the plutonium-239 for use in fresh fuel assemblies. However, the reprocessing of spent fuel from commercial nuclear reactors is not currently being done in the United States—although it is done in other countries, such as France, Japan, and Russia.

If the uranium fuel only passes through a nuclear reactor once and the resulting spent fuel is not reprocessed, then nuclear engineers refer to the fuel cycle as an open (or once through) fuel cycle. If only the uranium-235 is recovered for recycled use in fresh reactor fuel, then engineers call the fuel cycle a partially closed fuel cycle. If both the uranium-235 and plutonium-239 in spent reactor fuel are recovered by chemical processing for use in the nuclear fuel cycle, then engineers refer to the cycle as being a completely closed fuel cycle. The nuclear power industry in Japan is pursuing the use of recovered and recycled plutonium-239 in light-water reactors in the form of a mixed oxide fuel (MOX). Commercial nuclear power industries in other countries are also exploring this option. The use of mixed oxide fuel in light-water reactors is still a matter of much technical, economic, and political debate in the United States.

Nuclear engineers commonly divide the front end of the nuclear fuel cycle into the following steps: uranium exploration, mining, milling, uranium conversion, enrichment, and fuel fabrication. The back end of the fuel cycle includes the following steps: interim spent fuel storage, reprocessing, and waste disposal.

After its operating cycle, nuclear engineers shut a reactor down for refueling. The spent fuel discharged at that time is highly radioactive due to an accumulation of fission products. The spent fuel is initially stored at the reactor site in a spent fuel cooling pool. The spent nuclear fuel is usually stored in water since the water provides both cooling and radiation shielding. The spent fuel continues to generates heat and emit ionizing radiation due to radioactive decay of the many fission products it contains. The absence of a permanent geologic repository for spent fuel has caused a significant accumulation of spent fuel at commercial nuclear power plants throughout the United States.

Spent fuel discharged from light-water reactors contains appreciable quantities of fissile material (namely uranium-235 and plutonium-239), fertile nuclides (namely, uranium-238), and many other radioactive materials, including significant quantities of strontium-90 and cesium-137. These fissile and fertile materials can be chemically separated and recovered from the spent fuel. If economic and institutional conditions allow, the recovered uranium and plutonium can be recycled for use as nuclear fuel in commercial power plants. Although not practiced at present in the

United States, facilities in Europe and Japan reprocess spent fuel from nuclear electric utilities in Europe and Japan.

A current concern in the commercial nuclear power field is the safe disposal and isolation of either the spent fuel from reactors, or, if the reprocessing option is being used, the high-level wastes from reprocessing plants. These radioactive materials must be isolated from the biosphere for hundreds to thousands of years until their radioactive contents have diminished to a safe level.

Previous plans in the United States called upon the DOE to develop and operate the waste disposal system for spent fuel and high-level wastes. These plans also include the ultimate disposal of spent fuel and high-level wastes in solid form in licensed deep, stable geologic structures. The DOE's Waste Isolation Pilot Project (WIPP) near Carlsbad, New Mexico, stores transuranic waste. WIPP is an engineered (mined), deep geologic structure in an ancient salt bed. The other permanent geologic disposal facility for spent fuel and high-level waste that had been under consideration by the DOE was the Yucca Mountain site about 100 miles (160 km) northwest of Las Vegas, Nevada. On March 3, 2010, the DOE filed a motion with the Nuclear Regulatory Commission (NRC) to withdraw the license application for a high-level nuclear waste repository at Yucca Mountain. This action leaves the nation without a permanent solution to the high-level nuclear waste disposal issue.

By the end of the 20th century, commercial nuclear power plants in the United States had accumulated more than 39,000 metric tons of spent fuel. Nuclear energy analysts anticipate that this amount will double by the year 2035. Once the spent fuel and high-level waste (from DOE facilities) arrives by truck or rail at some future licensed and approved geologic repository, the waste will be removed from the special shipping containers and placed in long-lived waste packages for disposal. The waste would then be then carried into the underground repository by rail cars; placed on supports in the tunnels; and monitored until the repository is finally closed and sealed.

In the early 1980s, the need for alternative storage of spent fuel increased, as the spent fuel cooling pools at many commercial reactor sites in the United States began to reach capacity. Engineers at various utility companies investigated interim storage options, such as dry cask storage.

Dry cask storage allows spent fuel that has already been cooled in the spent fuel pool for at least one year to be surrounded by inert gas in a container called a cask. The casks are typically steel cylinders that are either welded or bolted closed. The steel cylinder provides a leak-tight

Spent fuel dry cask storage located on-site at a commercial nuclear power plant *(NRC)*

containment of spent fuel. Each cylinder is surrounded by additional steel, concrete, and other materials to provide adequate radiation shielding. Some cask designs are used for both interim (on-site) storage as well as transportation. In some dry storage cask system designs, the steel cylinders containing the spent fuel are placed vertically; other designs arrange the casks horizontally. For additional protection at an interim storage location, engineers can also place dry storage casks inside in a secured concrete bunker.

The NRC licensed the first dry cask storage installation in 1986 at the Surry Nuclear Plant in Virginia. Today, spent fuel is now being stored in dry cask systems at a number of commercial power plant sites across the United States. There is also an interim storage facility at the Idaho National Environmental and Engineering Laboratory.

NUCLEAR FUSION

In nuclear fusion, lighter atomic nuclei are joined together, or fused, to form a heavier nucleus. For example, the fusion of deuterium (D) with tritium (T) results in the formation of a helium nucleus and a neutron. Because the total mass of the fusion products is less than the total mass of the reactants (that is, the original deuterium and tritium nuclei), a tiny amount of mass has disappeared, and the equivalent amount of energy (about 17.6 MeV) is released in accordance with Einstein's mass-energy equivalence formula. This fusion energy then appears as the kinetic energy of the reaction products. When isotopes of elements lighter than iron fuse together, some energy is liberated. However, energy must be added to any fusion reaction involving elements heavier than iron.

The Sun is humankind's oldest source of energy, the very mainstay of all terrestrial life. The energy of the Sun and other stars comes from thermonuclear fusion reactions. Fusion reactions brought about by means of very high temperatures are called thermonuclear reactions. The actual temperature required to join, or fuse, two atomic nuclei depends on the nuclei and the particular fusion reaction involved. The two nuclei being joined must have enough energy to overcome the Coulomb (or like-electric-charge) repulsion. In stellar interiors, fusion occurs at temperatures of tens of millions of rankines (R) (kelvins [K]). Although many different nuclear fusion reactions occur in the Sun and other stars, only a few such reactions are of practical value for the potential production of energy on Earth. As they try to develop useful controlled thermonuclear reactions (CTRs), scientists consider reaction temperatures of 90 to 180 million rankines (50 million to 100 million kelvins) necessary.

At present, there are immense technical challenges preventing the effective use of controlled fusion as a terrestrial energy source. The key problem is that the fusion gas mixture must be heated to millions of degrees and held together for a long enough period of time for the fusion reaction to occur. For example, a deuterium-tritium (D-T) gas mixture must be heated to at least 90 million R (50 million K)—and scientists consider this reaction to be the easiest controlled fusion reaction to achieve. At 90 million R (50 million K), any physical material used to confine these fusion gases would disintegrate, and the vaporized wall materials would then cool the fusion gas mixture, quenching the reaction.

There are three general approaches to confining these hot fusion gases, or plasmas: gravitational confinement, magnetic confinement, and inertial confinement. Because of their large masses, the Sun and other stars are able to hold the reacting fusion gases together by gravitational confinement. Interior temperatures in stars reach tens of millions of degrees and use complete thermonuclear fusion cycles to generate their vast quantities of energy. For main sequence stars like or cooler than the Sun (with core temperatures of about 27 million R [15 million K] or less), the proton-proton cycle is believed to be the principal energy-liberating mechanism. The overall effect of the proton-proton stellar fusion cycle is the conversion of hydrogen into helium. Stars hotter than the Sun (those with interior temperatures greater than 27 million R [15 million K]) release energy through the carbon cycle. The overall effect of this thermonuclear cycle is again the conversion of hydrogen into helium, but this time with carbon (carbon-12 isotope) serving as a catalyst.

Here on Earth, nuclear scientists are attempting to achieve controlled fusion through two techniques: magnetic-confinement fusion (MCF) and inertial-confinement fusion (ICF). In magnetic confinement, strong magnetic fields are employed to bottle up, or hold, the intensely hot plasmas needed to make the various single-step fusion reactions occur, such as the D-T reaction. In the inertial-confinement approach, pulses of laser light, energetic electrons, or heavy ions are used to very rapidly compress and heat small spherical targets of fusion material. This rapid compression and heating of an ICF target allows the conditions supporting fusion to be reached in the interior of the pellet before it blows itself apart. Unlike nuclear fission, there are still many challenging technical issues to be resolved before scientists can manipulate nuclear fusion reactions that yield useful quantities of energy in a controlled and long-term manner.

The ITER Project is an international effort in magnetic confinement fusion being carried out by scientists from the United States, Russia, China, India, Japan, South Korea, and the European Union. The acronym ITER originally stood for *International Thermonuclear Experimental Reactor*; but now simply means "journey" or "direction" (from the Latin word *iter*), thereby avoiding any politically negative connotations in the word thermonuclear. The project's long-term goal is to develop the world's first full-scale, electricity-producing fusion power plant (ca. 2050). The ITER demonstration fusion reactor is now scheduled for completion in 2018 at a site in southern France. The system is being designed to produce 500 MW of output power for 50 MW of input power.

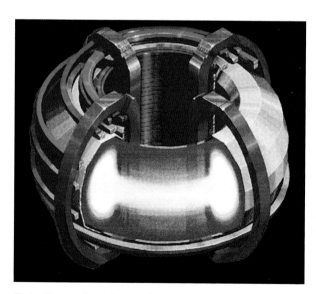

This illustration shows the basic magnetic-confinement fusion (MCF) concept. An intense magnetic field confines very hot deuterium and tritium plasma at the required density and temperature to yield useful quantities of output energy. *(DOE)*

Detonated from a barge near Bikini Atoll on March 26, 1954, the 11-megaton yield Romeo Event was part of Operation Castle—a series of high-yield thermonuclear weapon tests. *(National Nuclear Security Administration/Nevada Site Office)*

In sharp contrast to the ongoing scientific attempts at controlled nuclear fusion for power applications, since the early 1950s nuclear weapon designers have been able to harness (however briefly) certain fusion reactions in advanced nuclear weapon systems called thermonuclear devices. In typical two-stage nuclear explosives, the energy of a fission device is used to create the conditions necessary to achieve (for a brief moment) a significant number of fusion reactions of either the deuterium-tritium (D-T) or deuterium-deuterium (D-D) kind. Nuclear weapon scientists have designed and tested very powerful thermonuclear weapons that have total explosive yields in the multi-megaton (MT) range.

CHAPTER 9

Renewable Energy Resources

This chapter discusses renewable energy sources. Scientists and engineers point out that renewable energy sources can be replenished. The five renewable energy sources that are used most frequently in the United States are hydroelectric power, geothermal energy, wind power, solar energy, and biomass. According to the *2010 Electric Power Annual* published by the U.S. Energy Information Administration (EIA), 6 percent of the nation's 4.1 trillion kilowatt-hours of electricity came from hydroelectric power, while all the other renewable sources contributed a total of only 3 percent.

About 150 years ago, renewable energy in the form of wood (a biomass) supplied up to 90 percent of the nation's energy needs. As the use of coal, petroleum, and natural gas expanded, the United States became less reliant on wood as an energy source. Energy planners are now examining the increased use of renewable energy sources to reduce the nation's dependence on fossil fuels. An additional benefit of this strategy is that non-biomass renewable energy sources (such as hydroelectric, geothermal, wind, and solar) do not directly emit greenhouse gases. One of the major disadvantages of renewable sources is the fact that they are not always available when needed. Cloudy days can reduce the performance of solar energy systems; droughts can limit the amount of flowing water available for hydroelectric generation; and calm days cause wind power systems to remain idle.

HYDROELECTRIC POWER

Since ancient times, people have constructed dams for a variety of reasons. These include the impoundment of water in reservoirs for drinking and irrigation, the creation of artificial lakes (behind the dam) for recreation, the diversion or control of floodwaters, and the harnessing of flowing water's energy.

Engineers define hydropower as the use of flowing water to power machinery or to generate electricity (hydroelectric power). Contrary to popular belief, not all dams are used to produce electric power. Of the approximately 80,000 dams found in the United States, only 2,400 generate electricity. The vast majority of American dams support other uses, such as creating reservoirs for drinking water supplies and irrigation, flood control, recreation, and the production of human-made agricultural ponds for livestock.

In the Greco-Roman era, waterwheels of various designs and complexities served two basic functions: to grind grain or to lift (pump) water in support of irrigation practices. Following the Dark Ages, waterwheels reappeared throughout medieval Europe, many associated with monasteries. Clever arrangements of wooden gears allowed hydropower to support sawing operations (sawmills) and other labor-intensive mechanical activities.

At the start of the first Industrial Revolution, many of the late 18th-century British textile mills were located near dependable sources of flowing water. The arrival of commercially viable steam engines made the location of new factories independent of flowing water sources. Despite the technology transition, waterwheel-powered mills remained important in Great Britain, on the European continent, and in North America. For example, the first paper mill constructed on the Pacific Coast of the United States was located in Lagunitas, Marin County, California. Built in 1856, this paper mill originally used water power but later switched to steam.

Toward the end of the 19th century, the marriage of water power and electricity took place and changed the world. One historic example will highlight the intellectual excitement that accompanied the arrival of this important technical milestone. The visionary American entrepreneur George Westinghouse helped create the modern electric power industry by financially supporting Nikola Tesla and his development of alternating current (AC) generators, motors, and transformers. In 1895, Westinghouse's company won a coveted contract to use Niagara Falls to generate electricity and to deliver the hydropower-generated (AC) electricity to the city of Buffalo, New York, a city located 22 miles (35 km) away.

There are three basic types of hydroelectric facilities: impoundment, diversion, and pumped storage. The most common type of hydroelectric power plant is an impoundment facility, like Hoover Dam. As shown in the illustration on page 166, an impoundment dam stores river water in a reservoir. As water is released from the reservoir through a penstock, it spins a turbine-generator system, producing electricity. The electricity is then transmitted away from the dam by high-voltage, long-distance power lines. Dam operators can release water according to electric power demands or at a controlled rate to maintain a specific water-level range in

HOOVER DAM

Spanning the Colorado River between Arizona and Nevada, about 30 miles (48 km) southeast of Las Vegas, is an engineering wonder of the modern world known as Hoover Dam. Named after the 31st American president, Herbert Hoover (1874–1964), this massive concrete arch-gravity dam impounds an enormous quantity of water, using gravity and the forces generated by its horizontal arch design.

Hoover Dam is one of the world's largest dams. The U.S. Bureau of Reclamation began construction in 1931, completed it in 1936, and continues to operate the facility. Workers used more than 3.25 million cubic yards (2.46×10^6 m³) of concrete to build the dam, which rises 726.4 ft (221.4 m) from the foundation rock to the roadway on its crest. The maximum hydrostatic (water) pressure at the base of the dam is 312.5 psi (2.155 MPa).

Lake Mead is the reservoir created by the Hoover Dam. With a surface area of approximately 250 sq mi (640 km²), it is the largest reservoir in the United States and contains (when full) 28.5×10^6 acre-feet (35.2 km³) of water. Water impounded by Hoover Dam serves municipal drinking water needs, supports irrigation-based agriculture in California, Arizona, and Mexico, and generates electric power. The dam has an installed hydroelectric generating capacity of 2,080 megawatts (MW). On average, Hoover Dam generates about 4 billion kilowatt-hours of electricity each year for use in Nevada, Arizona, and California, enough to serve 1.3 million people. The dam supports flood control, helps regulate the Colorado River, and creates the Lake Mead National Recreation Area.

the reservoir. The turbine generators in a typical hydroelectric dam are quite large.

In a diversion hydroelectric facility, engineers channel just a portion of the river's flow through a canal or penstock. Since a dam is typically not required, engineers sometimes refer to this type of hydroelectric plant as a run-of-river facility. Recognizing that water is much easier to store than electricity, engineer devised another type of hydroelectric plant, called the pumped storage plant. During periods of light electric power demand, the facility uses some of its surplus generating capacity to pump water from a lower reservoir back into the upper reservoir. During periods of increased electric demand, the previously pumped water is released through penstocks to turbine-generator systems to produce electricity.

Aerial view of Hoover Dam and a portion of its large reservoir, Lake Mead *(USDA)*

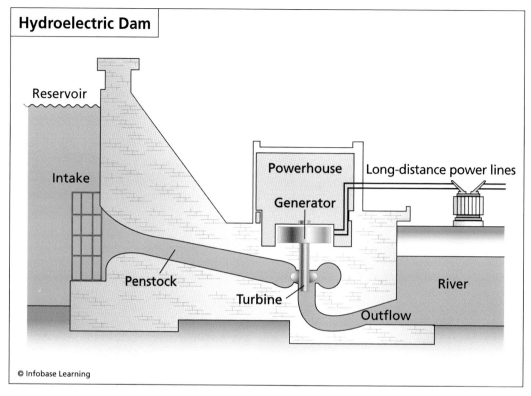

Hydroelectric Dam

Reservoir

Intake

Penstock

Turbine

Powerhouse

Generator

Long-distance power lines

River

Outflow

© Infobase Learning

Major components of a typical impoundment hydroelectric power plant *(TVA)*

This two-reservoir system represents a convenient way to store energy in the form of water (at a higher gravitational potential) until needed to satisfy peak electric loads.

Energy experts regard hydroelectric power generation as a nonpolluting, renewable source of energy. While the facility does not release air pollution or put toxic materials into the water, there are still some potential adverse environmental effects that need to be assessed. The construction of large dams restricts the flow of natural, wild rivers and blocks the migration of fish (like salmon) that must swim upstream to their spawning grounds. One solution that is partially successful is the use of properly designed fish ladders. Another environmental consequence of hydroelectric generation is that the reservoir created behind the dam floods the land, removing it from use for agriculture or for habitation by people or wildlife. Finally, a dam failure can release a flood of water downstream. Engineers and environmental planners carefully evaluate the major benefits and risks associated with each new hydroelectric project.

GEOTHERMAL ENERGY

The term *geothermal* refers to the heat associated with Earth's interior. Scientists regard the total amount of heat available within the Earth as enormous, so energy systems that use geothermal energy are treated as renewable energy systems.

Some applications of geothermal energy take advantage of shallow pockets of magma that heat groundwater. Water at elevated temperatures from hot springs has been used since ancient times for bathing, heating, and food preparation. Similarly, modern direct use geothermal energy systems and district heating systems employ hot water from springs near the surface. Bathing in hot springs remains very popular for many people. Where geographically feasible, engineers pipe the water from hot springs directly into buildings for space heat and into industrial facilities for use as process heat. A modern geothermal energy district heating system provides heat for 95 percent of the buildings in the city of Reykjavik, Iceland. The most common industrial application of geothermal energy is dehydration, the drying of fruit or vegetable products.

Geothermal energy can also be used for electric power generation. The United States now generates more geothermal electricity than any other country in the world; but the amount of electricity produced is less than 0.5 percent of all the electricity generated in the United States. The vast majority of geothermal electricity produced in the United States is generated by power plants in the state of California.

There are three basic types of geothermal power plant systems used to convert hydrothermal fluids into electricity: the dry steam power plant, the flash steam power plant, and the binary-cycle power plant. This section briefly describes the features and operational principles of each system. The type of conversion system depends on the thermodynamic state of the naturally heated working fluid (that is, water or steam), especially the fluid's temperature as it leaves the geothermal source and enters the power-generating equipment.

Dry steam power plants were the first type used for geothermal electric power generation. The first plant appeared in 1904 in Lardarello, Italy. The operational principle remains technically sound, and the approach to electric power generation still remains effective if suitable sources of good-quality, dry geothermal steam are available.

This type of plant uses hydrothermal fluids that are primarily steam. Wells take the steam from the geothermal source and feed it directly into a turbine, which drives the generator that produces electricity. The use of naturally heated steam in this type of power plant eliminates the need to burn fossil fuels to generate electricity. Dry steam power plants are used at The Geysers in northern California, the world's largest single source of geothermal power. The generating plants at The Geysers emit only spent (used) steam and very minor amounts of the gases normally associated with volcanic regions.

If the naturally heated geothermal fluid is above 360°F (182°C), engineers can use the fluid in a flash steam power plant to generate electricity. The principle of operation involves extracting the superheated fluid from the ground and spraying it into a large tank that is maintained at a much lower pressure than the fluid. When the high-pressure, heated fluid is sprayed into this low-pressure tank, some (if not all) the fluid flashes, or rapidly vaporizes. Engineers then use the vapor to spin a turbine, which drives an electric generator. In some flash steam power plant designs, engineers send any remaining hot liquid in the first rapid expansion tank into a second low-pressure tank to more thoroughly extract any remaining geothermal energy in the working fluid.

Most geothermal areas contain moderate-temperature (below 360°F [182°C]) water. Such moderate-temperature water is not really suitable for use in either dry steam power plants or flash steam power plants. So engineers have designed a binary-cycle power plant to more efficiently extract the geothermal energy from such geothermal sites. In a binary-cycle power plant, a secondary (or binary) working fluid extracts energy from the naturally heated water and then becomes a vapor that travels in a closed cycle through the turbine that spins the generator. The working fluid in the secondary (or binary), closed-cycle loop has favorable thermodynamic properties, allowing it to readily vaporize at moderate temperatures, pass through the turbine, and then cool back into a liquid by passing through an air-cooled condenser (heat rejection system). The hydrocarbon, butane (C_4H_{10}), is an example of a candidate binary working fluid for this type of geothermal power system. By design, the secondary (or binary) fluid circulates in a closed loop and does not get released into the atmosphere. As the moderate-temperature hot water passes through the binary-cycle heat exchanger it yields much of its energy to the binary-cycle working fluid and then travels as cooler water back to the underground geothermal reservoir.

HARVESTING THE WIND

People have harvested the power of the wind for centuries. Early windmills, looking like large paddle wheels, first appeared in the Middle East

This large modern wind turbine is of the type linked to bat mortality. The species of bats most susceptible to dying all roost in trees, leading some scientists to speculate that these animals may be visually mistaking the wind turbines for the trees in which they roost. *(USGS)*

more than a millennium ago. Some 200 year later, windmills began to appear in Europe. The primary purpose of these early wind machines was to grind grain. Centuries later, the people of the Netherlands improved the design of the basic windmill. They gave it propeller-type blades with which to extract energy from the winds. The blades of the windmill were connected to a drive shaft, which operated pumps. In past centuries, the people of Holland used such windmills to pump water from polders, as part of a major national effort to reclaim additional land from the North Sea.

Today, engineers are designing modern wind machines. These machines, called wind turbines, transform the kinetic energy of naturally flowing air into rotational mechanical energy suitable for the generation of electric power. Most modern wind machines are the horizontal-axis type. Horizontal-axis wind machines are typically very tall (about the height of a 20-story building) and have three blades that span 200 feet (61 m) across or more. As the blades of the wind turbine spin, they provide rotary mechanical energy (via a gearbox) to an electric generator.

Although wind energy is a useful renewable energy source, wind turbines cannot produce electricity when the wind is not blowing. Several environmental issues also arise, including noise (from the rotating mechanical equipment), aesthetics, and avian and bat mortality rates, as many of these flying creatures strike rotating blades while attempting to transit a wind farm.

SOLAR ENERGY SYSTEMS

The Sun is humans' parent star and the massive, luminous celestial object about which all other bodies in the solar system revolve. It provides the light and warmth upon which almost all terrestrial life depend. Scientists define solar radiation as the total electromagnetic (EM) radiation emitted by the Sun. They treat the Sun as a blackbody emitter, radiating energy at a surface temperature of 10,386 R (5,770 K). Approximately 99.9 percent of this radiated energy lies within the wavelength interval from 0.15 to 4.0 micrometers (μm), with some 50 percent lying within the visible portion of the EM spectrum (namely in the 0.4 to 0.7 μm wavelength interval), and most of the remaining solar energy radiated in the near-infrared portion of the spectrum. The region of the solar spectrum between 0.15 and 0.4 μm wavelength corresponds to the ultraviolet portion of the EM spectrum. (See chapter 7.)

Scientists define the solar constant as the total amount of the Sun's radiant energy that normally crosses perpendicular to a unit area at the top of Earth's atmosphere, when the planet is at one astronomical unit (93 million miles [149.6 million km]) from the Sun. The solar constant has a value of 436 Btu/h-ft^2 (1,375 W/m^2). Most of the Sun's radiant energy lies in the visible portion of the EM spectrum and has a peak value near 0.45 μm wavelength. The American astronomer and aeronautics pioneer Samuel Pierpont Langley (1834–1906) developed instruments (called bolometers) to accurately measure the intensity and spectral features of the solar radiation falling on Earth. His invention of the bolometer and his careful measurements of infrared radiation from the Sun foreshadowed modern concerns about climate change and the search for links between solar and terrestrial phenomena. Following in Langley's footsteps, scientists use the term *insolation* to quantify the amount of solar energy that falls on a planetary body. They typically express insolation as radiant flux per unit area per unit time. The amount of insolation at the top of Earth's atmosphere is simply the solar constant. However, the amount of insolation received by a specific area on Earth's surface depends not only upon the value of the solar constant, but also the latitude of the area of interest, the transparency of the atmosphere, and the season of the year. Insolation values determine whether a particular surface area is suitable for solar energy system applications. For example, cloud-free, lower-latitude regions generally have higher insolation values than traditionally overcast, higher latitude regions.

In December 2007, engineers at Nellis Air Force Base, Nevada (near Las Vegas), completed the largest solar photovoltaic (PV) system in North America. The solar field (partially shown here) consists of more than 72,000 solar panels and can generate more than 30 million kilowatt-hours of electricity annually. Located on 140 acres (57 ha) of capped landfill at the base, the huge PV array contains nearly 6 million solar cells. *(U.S. Air Force)*

Solar photovoltaic (PV) conversion involves the direct conversion of sunlight into electricity by means of the photovoltaic effect. Scientists call a single PV converter cell a solar cell, while they call a combination of cells designed to increase the electric power output a solar array or a solar panel.

Since 1958, engineers have used solar cells to provide electric power for a wide variety of spacecraft. Aerospace industry experience with photovoltaic materials serves as a technical heritage upon which other engineers can draw as they design and develop less expensive solar cells for use in electric energy generation projects here on Earth.

The typical solar cell is made of a combination of n-type (negative) and p-type (positive) semiconductor materials, generally silicon. When this combination of materials is exposed to sunlight, some of the incident electromagnetic radiation removes bound electrons from the semiconductor material atoms, thereby producing free electrons. A hole (positive charge) is left at each location from which a bound electron has been removed. Consequently, an equal number of free electrons and holes are formed. An electrical barrier at the p-n junction causes the newly created free electrons near the barrier to migrate farther into the n-type material and the matching holes to migrate farther into the p-type material.

If electrical contacts are made with these n-type and p-type materials and these contacts connected through an external load (conductor), the free electrons will flow from the n-type material to the p-type material. Upon reaching the p-type material, the free electrons will enter existing holes and once again become bound electrons. The flow of free electrons through the external conductor represents an electric current that will continue as long as more free electrons and holes are being created by exposure of the solar cell to sunlight. This is the general principle of solar photovoltaic conversion.

The performance of a photovoltaic array is dependent upon sunlight. When photons strike a photovoltaic cell, they may be reflected, pass right through, or be absorbed. Only the absorbed photons provide energy to generate electricity. Climate conditions (such as clouds or fog) have a significant effect on the amount of solar energy received by a photovoltaic array and, in turn, its performance. Most modern modules are about 10

(opposite) This diagram illustrates the photovoltaic energy conversion principle. When exposed to sunlight, a sandwich of special semiconductor materials produces electricity directly. *(adapted from State of Texas/ Energy Education Program)*

percent efficient in converting sunlight. Further research is being conducted to raise this efficiency to 20 percent.

Photovoltaic cells, like batteries, generate direct current (DC), which
is generally used for small loads (electronic equipment). When DC from

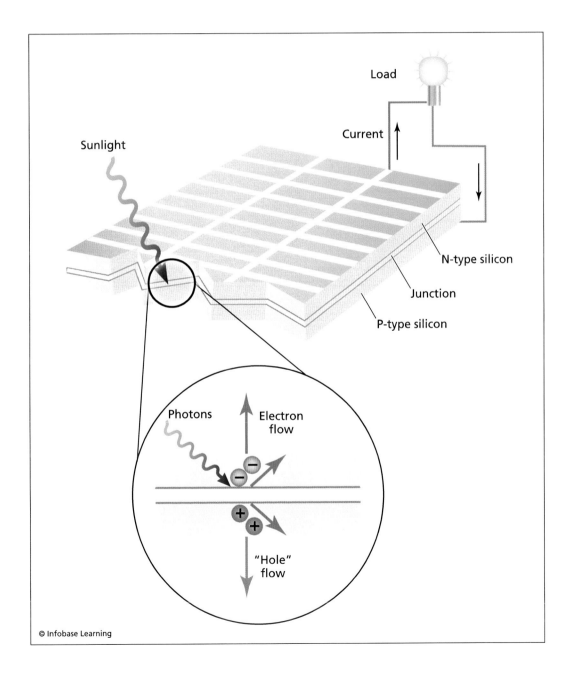

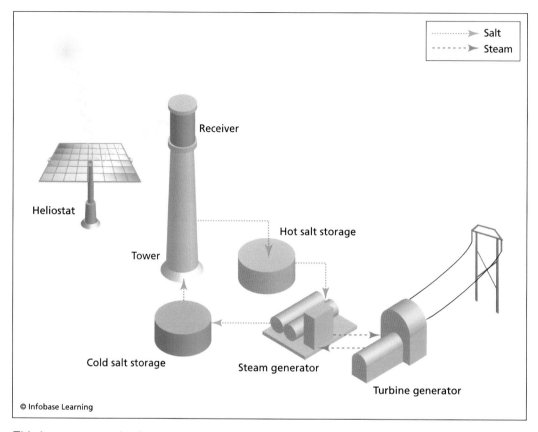

This is a conceptual solar power tower using an array of Sun-tracking heliostats (only one of hundreds shown here) to concentrate incoming sunlight onto a central, tower-mounted receiver. *(adapted from DOE artwork)*

photovoltaic cells is used for commercial applications or sold to electric utilities for use in the electric grid, it must be converted to alternating current (AC) by means of inverters—solid state devices that convert DC power to AC.

Solar thermal conversion involves the conversion of sunlight (solar energy) into electricity by means of a thermodynamic cycle involving a heat engine. In general, this type of system concentrates and focuses incoming sunlight in such a manner as to heat a working fluid. The solar-heated working fluid undergoes a series of changes in its thermodynamic state, during which mechanical energy (work) is extracted from the hot fluid. The mechanical (work) is then used to spin a generator to produce electricity. Thermodynamic cycles that can be used in solar thermal conversion include the Brayton, Rankine, and Stirling cycles.

The illustration on page 174 shows the major components of a conceptual power tower approach to solar thermal conversion. The concept employs molten nitrate salt as the working fluid. Molten nitrate salt is a clear liquid with properties similar to those of water, except this salt has a melting point of 464°F (240°C). Hundreds of heliostats (only one is shown in the figure) track the Sun and concentrate incoming solar energy on the receiver. Cold working fluid is pumped from a large, molten salt storage tank to this receiver, where it is heated in tubes to a temperature of about 1,049°F (565°C). The heated molten salt is then sent to a second large storage tank, where it remains until it is removed for electric power generation activities. During electric power generation, the hot molten salt passes through a steam generator, surrendering its heat to transform water, which is flowing through separate tubes, into the steam. The steam then spins a high-efficiency turbine-generator. After passing through the steam generator, the molten-salt working fluid, has cooled to about 545°F (285°C). In the final step in this cyclic process, the working fluid is pumped back to the cold salt storage tank. It remains there, until pumped back again through the receiver in the solar power tower.

The advantage of this type of solar thermal conversion system is that it can generate electricity during peak demands (usually at night), even though the Sun is no longer shinning on the local region. Unlike most other solar energy systems, which can produce power only when the Sun is shining, the molten-salt power tower can generate electricity not only on sunny days but also during cloudy periods or at night. Depending on the amount of heated molten salt in storage, this type of system can provide electric power for up to 12 hours after sunset. Engineers suggest that this type of power plant would typically generate a maximum of about 10 megawatts of electricity.

BIOMASS ENERGY

Scientists define biomass as organic material made from plants and animals. Biomass contains stored solar energy. A green plant has the unique ability to harvest solar energy. Through the process of photosynthesis, plants convert radiant energy from the Sun into chemical energy in the form of glucose ($C_6H_{12}O_6$) or sugar. Chemists describe the photosynthesis of glucose with the following reaction equation: $6\ CO_2 + 6\ H_2O$ + sunlight (radiant energy) $\rightarrow C_6 H_{12}O_6 + 6\ O_2$. In addition to storing incoming solar energy as carbohydrate molecules, green plants take in carbon dioxide

(CO$_2$) and release oxygen (O$_2$). Plant cells can convert the carbohydrate molecules formed during photosynthesis into fat molecules, and with the right combination of inorganic nutrients, into protein molecules.

Animals obtain their chemical energy by eating plants or by eating other animals, including plant-eating species (herbivores) and various meat-eaters (carnivores). The three major types of foods for animals are carbohydrates, proteins, and fats. The process of metabolism represents the entire sequence of chemical activities and processes that keep cells alive. Scientists define catabolism as the degradation of molecules to provide energy; they define anabolism as the metabolic process in which living cells use energy to build up (synthesize) molecules.

Included within the broad category of biomass fuels are wood, crops, manure, and certain types of municipal waste (garbage). Scientists regard biomass as a renewable energy source, because people can grow more trees, crops (like corn), and bioenergy plants (like switchgrass), and in the process of day-to-day living people always generate garbage.

The chemical energy stored in biomass is released as heat during combustion. There is also a release of carbon dioxide (CO$_2$) when biomass is burned. For most of human history, people burned wood for warmth, light, and to cook food. People have also combusted wood (and other biomasses) to produce steam to power machinery or to spin turbine generators to produce electricity.

Biomasses such as corn and sugarcane, can also be fermented to produce biofuels, such as ethanol. Chemists refer to ethanol as ethyl alcohol (CH$_3$CH$_2$OH) or grain alcohol. Ethanol was one of the first fuels used by people driving early internal combustion engine automobiles. In an effort to reduce national

This is an artist's rendering of switchgrass, the tall native grass of the North American prairie. This fast-growing plant can reach a height of 10 feet (three m) at maturity. Once food for millions of bison, scientists view switchgrass as a bioenergy crop. Research suggests that an acre of switchgrass, once harvested and processed, yields about 1,150 gallons (4,369 L) of ethanol each year. As an added bonus, any portions of the biomass not converted to ethanol can be combusted to generate electricity. *(DOE/Genomic Science Program)*

dependence on foreign petroleum supplies, American engineers are revisiting the use of ethanol as a domestically produced, renewable transportation fuel. Today, most ethanol produced in the United States is distilled from corn. About 99 percent of the current corn-based ethanol production (primarily in the American Midwest) goes into making E10 (gasohol), a commonly encountered automotive fuel that contains 10 percent ethanol and 90 percent gasoline. Any modern automobile gasoline engine can operate with E10 fuel. The remaining 1 percent of American ethanol production goes into E85, a special automotive fuel blend containing 85 percent ethanol and 15 percent gasoline. Only specially engineered automobiles, called flexible fuel vehicles (FFVs), can operate with E85 fuel. Engineers have designed FFVs so they can operate with any mixture of ethanol and gasoline, up to and including E85.

Another biofuel is biodiesel, a renewable fuel made from vegetable oils, fats, or greases—including recycled restaurant grease. Biodiesel fuel is safe, biodegradable, and compatible with modern diesel engines. Pure biodiesel fuel is called B100, but it can also be blended with petroleum diesel in ratios (biofuel-to-petroleum fuel) of 2 percent (B2), 5 percent (B5), or 20 percent (B20). Biodiesel fuels can be used in regular diesel engines without making any modifications to these engines. However, B100 is a solvent and can cause rubber and other components to fail in older vehicles. Low-level biodiesel blends, like B2 through B5, are very popular in the trucking industry, because biodiesel has excellent lubricating properties that prove beneficial for engine performance.

There are many other interesting topics involving biomass that exceed the scope and limits of this chapter. These include recovering energy from municipal solid waste (garbage), collecting biogas (methane) at landfills, and the accumulation of animal wastes in digesters to generate methane gas for electric power production.

CHAPTER 10

Hydrogen—the Fuel of Tomorrow

> *I believe that water will one day be employed as fuel, that hydrogen and oxygen which constitute it, used singly or together, will furnish an inexhaustible source of heat and light, of an intensity which coal is not capable.*
>
> —Jules Verne, *The Mysterious Island* (1874)

This chapter discusses the potential role hydrogen can play in the global energy infrastructure of the 21st century. Special attention is given to use of hydrogen fuel cells in transportation. One future scenario involves the extensive use of renewable energy sources to produce hydrogen during periods of low electricity demand. Utility companies can then use the stored hydrogen in fuel cells or as a combustible fuel in heat engines to produce electricity when the demand for electric power peaks.

PROPERTIES AND USES OF HYDROGEN

Scientists had been producing hydrogen for years, before it was recognized as an element. Written records suggest that the Irish-British scientist Robert Boyle (1627–91) produced hydrogen gas as early as 1671, while experimenting with iron and acids. Hydrogen was first recognized as a distinct element in 1766 by the British scientist Henry Cavendish (1731–1810) and later named by the French chemist Antoine Lavoisier. Starting in 1783, Lavoisier conducted a series of brilliant combustion experiments, during which he demonstrated that water was composed of only hydrogen and oxygen.

Composed of a single proton and a single electron, hydrogen is the simplest and most abundant element in the universe. It is estimated that 90 percent of the visible universe is composed of hydrogen. Hydrogen is the basic thermonuclear fuel that most stars burn to release energy. The same process, known as fusion, is being studied as a possible power source for use on Earth. Astronomers anticipate that the Sun's supply of hydrogen will last another 5 billion years.

Hydrogen has three common isotopes. The simplest isotope, called protium, is just ordinary hydrogen. The second, a stable isotope called deuterium, was discovered in 1932. The third isotope, called tritium, was discovered in 1934. Tritium is radioactive and has a half-life of 12.5 years.

Although hydrogen is the most abundant element in the universe, it is not found by itself as an element here on Earth. Hydrogen gas is so much lighter than air that it rises fast and is quickly ejected from the

This is one of the small fleet of hydrogen fuel-cell vehicles that roamed Selfridge Air National Guard Base, Michigan, as part of an extensive vehicle performance test program in 2009. For more than a year, technicians were also trained on how to maintain the engines and related components of hydrogen-powered fuel cell vehicles. *(U.S. Air Force)*

atmosphere. This is why hydrogen as a gas (H_2) is not found by itself on Earth. It is found only in compound form with other elements. Hydrogen combined with oxygen is water (H_2O). Hydrogen combined with carbon forms different compounds, including methane (CH_4), coal, and petroleum. Hydrogen is also found in all growing things—for example, biomass. It is also an abundant element in the Earth's crust.

The name hydrogen means water forming. This gaseous element has an atomic number of 1, an atomic weight of 1.00794, a melting point of −434.81°F (−259.34°C [13.81 K]), and a boiling point of −423.1781°F (−252.87°C [20.28 K]).

Hydrogen is a commercially important element. Large amounts of hydrogen are combined with nitrogen from the air to produce ammonia (NH_3) through a process called the Haber process. Hydrogen is also added to fats and oils, such as peanut oil, through a process called hydrogenation. Hydrogen combines with other elements to form numerous compounds. Some of the common ones are water (H_2O), ammonia (NH_3), methane (CH_4), table sugar ($C_{12}H_{22}O_{11}$), hydrogen peroxide (H_2O_2), and hydrochloric acid (HCl).

Liquid hydrogen is used in the study of superconductors and, when combined with liquid oxygen, makes an excellent rocket fuel. Liquid hydrogen also plays a role in the study and application of high-temperature superconductors. Energy engineers regard liquid hydrogen as an important energy carrier of the future.

An energy carrier is a substance or system that moves energy in a usable form from one place to another. Today, electricity is the most well-known energy carrier. People use electricity to transport the energy available in coal, uranium, or falling water from generation facilities to points of application like industrial sites and homes. An energy carrier makes the application of the energy content of the primary source more convenient. Liquid hydrogen can store energy until it is needed and can also move energy to a variety of places where it is needed.

Since hydrogen is not found as a free element on Earth, engineers have developed processes to separate it from other elements. At present, the two most common methods for producing hydrogen are steam reforming and electrolysis. The least expensive method is steam reforming, which accounts for about 95 percent of the hydrogen produced in the United States. In this process, hydrogen atoms are separated from the carbon atoms in methane (CH_4) molecules. Typically, chemical engineers employ an external source of hot gas to heat tubes in which a catalytic reaction takes place, converting steam (very hot H_2O) and lighter hydrocarbons, such as natural gas (methane [CH_4]), into hydrogen and carbon monoxide (CO).

In the other production process, called electrolysis, hydrogen atoms are harvested after water molecules (H_2O) are split. Electrolysis is currently quite costly, but engineers are examining future approaches to hydrogen production at large central plants or small local plants. Once liquid hydrogen becomes less expensive and more available, it can be used in transportation systems as an environmentally friendly, combustible fuel (since the combustion by-product is water). Hydrogen can also be used in fuel cells to generate electricity for vehicles, homes, and industrial sites.

As an extremely cold cryogen, proper equipment is needed to safely store and handle liquid hydrogen. Hydrogen gas is flammable, so care must also be exercised to avoid situations where gas leaks represent an explosive hazard. The flammable range of hydrogen in room temperature air at one atmosphere pressure is 4 to 75 percent by volume. The flame temperature of hydrogen in air is 3,713°F (2,045°C).

The American aerospace program has successfully handled liquid hydrogen for decades. Operational and safety procedures involving the space program's use of liquid hydrogen represent an important technical legacy. These experiences can assist the growth of a global hydrogen-based energy economy this century.

At liftoff, the external tank of a NASA space shuttle flight vehicle typically carried about 384,071 gallons (1.45×10^6 L) of liquid hydrogen. The hydrogen used at the Kennedy Space Center (KSC) was produced from natural gas by a steam-reforming process at a facility in New Orleans, Louisiana. The liquid hydrogen was then shipped to KSC in 13,000 gallon (49,200 L) mobile cryogenic tankers. The external tank also carried about 141,750 gallons (536,550 L) of liquid oxygen, produced locally at a plant near the launch site, which liquefies the air and separates the liquid oxygen.

HYDROGEN AS AN ALTERNATIVE TRANSPORTATION FUEL

Technical interest in hydrogen as an alternative transportation fuel arises from its clean-burning qualities, its potential for domestic production, and the higher conversion efficiency promised by fuel cell vehicles (FCVs).

Engineers point out that the energy in 2.2 pounds-mass (one kg) of hydrogen gas is about the same as the energy in one gallon (3.85 L) of gasoline. A light-duty FCV would have to store between 11 and 19 lbs (five and 13 kg) of hydrogen to achieve a driving range of about 300 miles (480 km) or more. Because hydrogen has a low volumetric energy density compared with fuels like gasoline, storing this mass of hydrogen on a vehicle using currently available storage technology results in a very large tank—usually larger than the tank of a typical automobile. Improved storage technologies are essential in making hydrogen suitable for use as an alternative fuel for transportation.

Some demonstration automobiles have been modified to use 100 percent hydrogen or hydrogen blended with compressed natural gas (CNG) as the fuel in their internal combustion engines (ICEs). These experiments are providing engineers important data about the feasibility of hydrogen-fueled ICE vehicles, as a potential alternative to FCVs. Along either technology pathway, hydrogen would become the dominant transportation fuel by the mid-21st century.

HYDROGEN FUEL CELLS

A fuel cell is a device that can directly convert the chemical energy in hydrogen and oxygen into electricity, with pure water and potentially useful heat as the only by-products. Hydrogen-powered fuel cells are not only pollution free but also are generally more efficient. In many fuel cells, hydrogen is the active material at the negative electrode and oxygen is the active material at the positive electrode. Since hydrogen and oxygen are often used as gases in fuel cells, a typical cell requires a solid electrical conductor to serve as the current collector and to provide a terminal at each electrode. This solid electrode material is generally porous and provides a large number of sites where the gas, the electrolyte, and the electrode are in contact. The cell's electrochemical reactions occur at these sites. The reactions are normally very slow, so engineers often include a catalyst in the electrode to expedite the process.

Scientists generally classify fuel cells by the type of electrolyte they use. The choice of electrolyte determines the kind of chemical reactions that take place in the cell, the type of catalysts required, the temperature range in which the cell operates, the fuel required, and other technical factors. The polymer electrolyte membrane (PEM) fuel cell delivers high power density and offers the advantage of low mass and small volume, when compared to other types of fuel cells. Also called proton exchange membrane cells, the PEM fuel cell employs a solid polymer as an electrolyte and porous carbon electrodes containing a platinum catalyst. Unlike other fuel cells which use corrosive liquids, PEM fuel cells require only hydrogen, oxygen (from the air), and water to operate. The cells are generally used with pure hydrogen supplied from storage tanks or with hydrogen extracted from a secondary fuel by means of an onboard reformer.

In the PEM fuel cell, hydrogen gas flows through channels to the anode, where a catalyst causes the hydrogen molecules to separate into constituent protons and electrons. The membrane allows only the protons to pass through it and travel toward the cathode. While the positively charged protons are conducted through the membrane to the other side of the cell, the stream of negatively charged electrons follows an external circuit to the cathode. The flow of electrons represents the direct current electricity that can be used to power a motor. On the other side of the PEM fuel cell, oxygen gas (normally drawn from outside air) flows through channels to the cathode. When the electrons return to the cell after passing through the external circuit, they react with oxygen and the protons (hydrogen nuclei) that have moved through the membrane at the cathode to form water. This union involves an exothermic reaction, liberating heat that can be used outside the fuel cell.

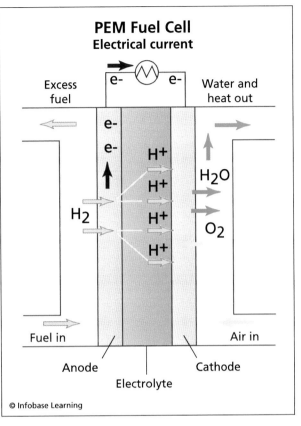

This is a simplified schematic diagram of a polymer electrolyte membrane (PEM) hydrogen fuel cell. *(adapted from DOE artwork)*

To increase the amount of electricity generated, engineers combine individual fuel cells in series to form stacks. Depending on the application, a fuel cell stack may contain only a few or as many as hundreds of individual cells layered together. The scalability feature of fuel cells make them quite suitable for a wide variety of applications requiring reliable electric power.

PEM fuel cells operate between 122°F and 212°F (50°C and 100°C), have power outputs ranging from less than one kilowatt-electric to 250 kilowatts-electric, and efficiencies in transportation systems applications from 53 to 58 percent and in stationary power generation systems from 25 to 35 percent. Engineers consider the PEM fuel cell as a promising device for use in light-duty transportation systems. They also consider this type of fuel cell suitable for backup power systems, portable power units, and small, distributed electric systems.

Engineers and scientists are investigating many other types of hydrogen fuel cells. These include the alkaline fuel cell (AFC), the phosphoric acid fuel cell (PAFC), the molten carbonate fuel cell (MCFC), and the solid oxide fuel cell (SOCF). The last two types of fuel cells appear capable of electric power generation in the megawatt-electric range, making them suitable candidates for electric utilities and for large distributed generation applications.

One of the best electrochemical catalysts is finely divided platinum or platinum-like metal deposited on or incorporated with the porous electrode material. The catalyst particles provide active sites at which the cell's electrochemical reactions can occur at a fairly rapid pace. The catalyst will gradually lose its ability during the overall operating lifetime of the fuel cell.

The electrochemical reaction in a hydrogen-oxygen fuel cell is the exact opposite of the electrolysis of water. Water electrolysis is the conversion of electrical energy into chemical energy in the form of hydrogen with oxygen as a useful by-product.

The number of fuel cells determines the voltage, just as the number of battery cells determines a battery's voltage. The concept of using an electrochemical process to store and generate electrical energy in a flashlight is familiar to most people. In a flashlight battery, the chemical energy is stored inside the cell, whereas in a fuel cell the chemical energy is stored outside the cell. Fuel cells will produce electricity as long as the fuel (hydrogen) and the oxidant (oxygen) are supplied.

The idea of deriving electrical energy directly from oxidizable fuels has attracted scientific attention for many years, because the method enables electric power generation without heat engine or mechanical devices. Some key advantages of fuel cells in electric energy generation are low noise and low pollution emissions, reliability and long operating life due to the absence of moving parts, and high energy conversion efficiency. Hydrogen-powered fuel cells are not only pollution-free but also have more than two times the efficiency of traditional combustion technologies.

A conventional combustion-based power plant typically produces electricity at thermodynamic efficiencies of between 33 to 35 percent, while fuel cell systems can generate electricity at efficiencies of up to 60 percent or higher. Under normal driving conditions, the gasoline engine in a conventional automobile is less than 20 percent efficient in converting the chemical energy in gasoline into the mechanical energy that moves the vehicle. Hydrogen fuel cell vehicles, which use electric motors, are much more energy efficient and use 40 to 60 percent of the hydrogen fuel's energy content.

Fuel cell vehicles can be fueled with pure hydrogen gas stored directly on the vehicle or extracted from a secondary fuel (such as methanol, ethanol, or natural gas) that carries hydrogen. If an FCV uses secondary fuels, these fuels must first be converted into hydrogen gas by an onboard device called a reformer. Fuel cell vehicles fueled with hydrogen emit no pollutants, only water and heat. FCVs that use a reformer to convert secondary fuels to hydrogen can produce a small quantity of air pollutants. Automotive engineers anticipate equipping FCVs with other energy-saving features, such as regenerative braking systems, which capture the energy lost during braking and store it in a large electric battery.

Because FCVs involve the introduction of a completely new vehicle propulsion system, as well as a new vehicle fueling infrastructure, there are many technical, political, and social issues that must be addressed. For example, automotive technicians must be retrained, thousands of gasoline stations must be transitioned to supplying hydrogen, and automobile manufacturers must retool and restructure assembly lines. One pressing social issue is what happens to the workers trapped in the transition from gasoline-powered automobiles to fuel cell vehicles. Some workers will adapt to the change and learn the new skills needed for employment; others will become obsolete and unemployable. The transition to hydrogen-fueled vehicles does not simply involve technical innovations and advances. Hydrogen offers great promise, but enlightened strategic planning by government and industry leaders is absolutely necessary to minimize workforce disruption, while maximizing the technical and environmental advantages of this exciting fuel. Once these issues are successfully addressed, fuel cell vehicles will become an integral part of the nation's transportation infrastructure in the 21st century.

In 2010, there were an estimated 200 to 300 hydrogen-fueled vehicles in the United States. Most of these vehicles are buses and automobiles powered by electric motors. They store hydrogen gas or liquid on board and convert the hydrogen into electricity for the motor using a fuel cell. Only a few of these vehicles burn the hydrogen directly (producing almost no pollution).

The present cost of fuel cell vehicles greatly exceeds that of conventional vehicles in large part due to the expense of producing fuel cells. However, hydrogen vehicles are starting to move from the laboratory to the road. Several are in use by a few state agencies and a few private entities. As of October 2010, there were about 60 hydrogen refueling stations in the United States, about half of which are located in California. There are several so-called "chicken and egg" questions that hydrogen-powered vehicle developers are working hard to solve, including who will buy hydrogen cars if there are no refueling stations? And who will pay to build a refueling station if there are no cars and customers? As these economic and technical issues get resolved over the next two decades, Earth's hydrogen-powered parent star will continue to shine upon the planet and make life possible.

VISION OF A GLOBAL HYDROGEN ECONOMY

This chapter opened with an excerpt from Jules Verne's *The Mysterious Island*. The French writer Jules Verne (1828–1905) created modern science fiction with many popular novels, such as his classic novel *From the Earth to the Moon,* written in 1865. Long-range planners and technical strategists often regard Verne as a technical prophet, a person who had an uncanny knack for successfully extrapolating contemporary technologies and scientific knowledge far into the future.

Hydrogen fuel cells (batteries) make electricity. They are very efficient but expensive to build. Small fuel cells can power electric cars. Large fuel cells can provide electricity in remote places with no power lines. Because of the high cost of building fuel cells, large hydrogen power plants will probably not be constructed for some time. However, fuel cells are being used in some places as a source of emergency power, from hospitals to wilderness locations. Portable fuel cells are being sold to provide longer power for laptop computers, cell phones, and military applications.

The illustration on page 187 suggests one compelling future scenario in which various renewable energy sources produce hydrogen during non-peak electricity demand periods. The hydrogen then becomes available

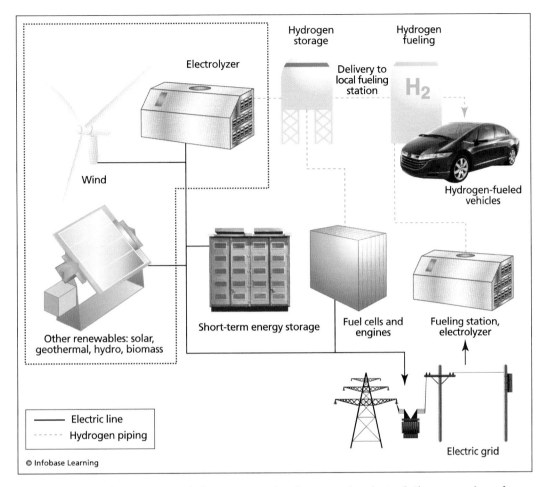

One conceptual future energy infrastructure that features the electrolytic conversion of renewable energy sources into hydrogen *(adapted from DOE/NREL artwork)*

to generate electricity through fuel cells or thermodynamic combustion cycles at times when the renewable sources are unavailable. Hydrogen, as an energy carrier, can also take renewable energy away from remote locations and deliver it (by means of gas pipelines or cryogenic-liquid tanker distribution systems) to other locations *where and when* it is needed to fuel vehicles or to generate electricity.

As with all energy source transitions (such as wood to coal and petroleum), the transition to hydrogen will take time and occur in phases. These phases will most likely be defined by a combination of technical advances and marketplace acceptance. Public acceptance of hydrogen will depend on safety perceptions, education, and the overall availability of hydrogen, especially in the transportation sector marketplace.

Today, strategic planners and technical futurists anticipate hydrogen assuming a much larger role in the global energy infrastructure in this century. Hydrogen represents a premier energy carrier and transportation sector fuel source. Whether hydrogen is produced by electrolytic or thermochemical methods, whether the initial source of energy is a coal-fired or nuclear plant, a solar power tower, a vast photovoltaic array, or a field of wind turbines, the resulting hydrogen fuel represents an energy carrier much more convenient and versatile than the coal, uranium, or even solar energy from which it derives.

Researchers around the world are aggressively exploring alternative ways of producing hydrogen. One very interesting idea comes from scientists at the Oak Ridge National Laboratory (ORNL), who are searching for a cost-effective means of making green hydrogen—hydrogen produced using algae, water, and sunlight.

CHAPTER 11

Conclusion

Civilization reflects the story of the human mind understanding the energy content of matter. The ability of human beings to relate the microscopic (atomic level) behavior of matter to readily observable macroscopic thermodynamic properties (such as density, pressure, and temperature) has transformed the world.

The origin and nature of matter and energy have perplexed people since the dawn of history. The story of energy and its relationship to matter starts with an event called the big bang. About 13.7 billion years ago, there was an incredibly powerful explosion that started the present universe. Before this energetic event, matter, energy, space, and time did not exist. All of these physical phenomena emerged from an unimaginably small, infinitely dense object, which scientists call the initial singularity. Immediately after the big bang, the intensely hot universe consisted of pure energy. As the universe expanded and cooled, matter began to form. At first, matter consisted of just an intensely energetic quark-gluon plasma. The further evolution of the universe involved the continual transformation of matter into energy, and vice versa.

It has taken the collective thinking of some of the very best minds in human history many millennia to recognize the key scientific fact that all matter in the observable universe consists of combinations of different atoms, drawn from a modest collection of about 100 different chemical elements. Scientists are now trying to understand and characterize an intriguing invisible form of matter called dark matter, as well as an equally mysterious phenomenon called dark energy. Detailed measurements of the cosmic microwave background revealed that the current contents of the universe include 4.6 percent atoms—the building blocks of stars, planets, and people. In contrast, dark matter comprises 23 percent of the universe. Finally, 72 percent of the current universe is composed of dark energy, which acts like a type of matter-repulsive, antigravity phenomenon.

In dealing with the energetic behavior of ordinary (or baryonic) matter, scientists have defined energy as an ability to do work. They have also divided energy into two basic categories: kinetic energy and potential energy. Kinetic energy is the energy contained or exhibited by matter in motion; while potential energy represents energy stored in a material body or system as a consequence of its position, state, or shape. Scientists often treat chemical energy, nuclear energy, electrical energy, and gravitational energy as various forms of potential energy.

Throughout most of human history, the process of understanding energy has been a very gradual one. The discovery and use of fire in distant prehistoric times represents the first major milestone. Then, several thousand years ago, people learned how to use fire to process metals and manufacture other new materials, like pottery and bricks. This discovery supported the rise of advanced civilizations all around the Mediterranean Sea and elsewhere.

One of the most important contributions of Western civilization to the human race is the emergence of modern science and the development of the scientific method. During the intellectually turbulent period of the 17th century, men of great genius such as Galileo Galilei and Sir Isaac Newton discovered important physical laws and constructed experimental techniques that helped people explain how matter and energy influenced the operation and behavior of the physical universe. The ability to quantify mechanical energy was a major step forward in understanding the behavior of the physical universe. Unraveling the mystery of heat was the next important milestone.

As scientists and engineers labored to make steam engines more efficient in the early 19th century, they began to explore the nature of heat and its relationship to mechanical work. They also used the scientific method to carefully investigate the thermal properties of matter, giving rise to thermodynamics. Steam not only powered the First Industrial Revolution, it also powered an amazing intellectual revolution, involving materials science, fluid mechanics, thermodynamics, and the rebirth of atomic theory.

This trend continued well into the Second Industrial Revolution, when great expansions within the oil, steel, chemical, and electrical industries occurred. Engineering developments in this period involved the expanded use of fossil fuel (especially coal and petroleum) as a prime energy source and of electricity as a revolutionary new energy carrier. Scientific discoveries during the Second Industrial Revolution also set the stage for the rise in modern physics, which (in turn) enables the next two technology-based revolutions in human history—the nuclear age and the digital (information technology) revolution.

The invention of the first electric battery (known as Volta's electric pile) stimulated the start of the Second Industrial Revolution. Soon laws governing the flow of electric currents were developed. Michael Faraday and Joseph Henry independently discovered the physical principles behind the electric generator and the electric motor. Later in the 19th century, James Clerk Maxwell presented a comprehensive set of equations that theoretically described electromagnetism. Nineteenth-century inventors such as Thomas Edison and Nikola Tesla applied electromagnetism to numerous new devices, which gave humankind surprising new power and comforts.

Early in the 20th century, Max Planck introduced quantum theory and Albert Einstein presented relativity theory. Their pioneering efforts founded modern physics and profoundly influenced many aspects of today's technology-dependent global civilization. Planck's pioneering work supported a more detailed, comprehensive understanding of matter and energy at the smallest imaginable scale. Specifically, Planck suggested that matter radiates energy not continuously, but rather in discrete chunks or quanta. When Einstein developed his theory of the photoelectric effect in 1905, he embraced Planck's notion of discrete energy packets by postulating that light itself is composed of energetic particles called photons. In addition, Einstein's special relativity included the revolutionary notion of the equivalence of mass and energy. His deceptively simple mass-energy formulation gave scientists the means of understanding the complex processes taking place in the atomic nucleus during fission, fusion, radioactive decay, and matter-antimatter annihilations. Energy was liberated in such nuclear reactions because equivalent amounts of matter disappeared. In the first half of the 20th century, other scientists explored the atomic nucleus and discovered way to harvest the incredible amounts of energy hidden within. Unlocking the energy within the atomic nucleus forever changed human destiny.

Energy, especially an abundant supply of electricity, empowers today's global civilization. A extended loss of electricity (that is, a blackout) for just a 24-hour period would propel most people living in an industrialized society back at least one century in quality of life. After three days without electricity, the electricity-deprived residents of a modern city would feel as though they had quite literally entered the neo–Dark Ages. Supplies of perishable foods would soon rot for lack of refrigeration, water supplies and sanitation system would fail to function because their electric pumps were inoperative, most forms of electronic communications and commerce would be severely disrupted, routine medical services would vanish, and nocturnal lighting would become totally dependent on candles, kerosene lanterns, and a rapidly dwindling supply of chemical batteries. To make matters worse, in a "de-electrified" urban environment, security and surveillance systems would be inoperative, mass transit rail systems would be paralyzed, high-rise apartment tenants would become virtual prisoners in their homes, and the streets below would be hopelessly snarled with unregulated motor vehicle traffic.

Although often taken for granted, the large-scale generation of electricity for use in modern society is the direct result of human ingenuity, scientific breakthroughs, and decades of excellence in engineering. Just like a few modern farmers can grow food each day to nourish an entire population, energy technologists work equally hard each day to make sure that people can satisfy their reasonable needs for electricity.

Scientists divide energy sources into two general groups: nonrenewable and renewable. Nonrenewable energy sources include the fossil fuels (coal, petroleum, and natural gas), as well the element uranium. Renewable energy resources include solar energy, geothermal energy, wind energy, hydropower, and biomass energy (such as wood). Scientists and engineers regard electricity as an energy carrier. Electricity is both a basic part of nature and one of humankind's most widely used forms of energy.

Like electricity, hydrogen is also considered a secondary source of energy and an energy carrier. The American aerospace program has successfully handled liquid hydrogen for decades. Operational and safety procedures involving the space program's use of liquid hydrogen represent an important technical legacy. These experiences can assist the growth of a global hydrogen-based energy economy this century.

Control of hydrogen, nature's most abundant element, represents the key to humankind's future on this planet. Hydrogen in fuel cells and as the combustible fuel for traditional thermodynamic power conversion represents an environmentally friendly approach to electric power generation and transportation. Harnessing controlled thermonuclear fusion offers incredible possibilities here on Earth and makes the entire solar system accessible to more intense exploration and ultimately human habitation. About 5 billion years ago, this solar system evolved from a huge cloud of hydrogen gas; now that same incredibly interesting element represents the energy key to the future of the human race.

Appendix

Scientists correlate the properties of the elements portrayed in the periodic table with their electron configurations. Since, in a neutral atom, the number of electrons equals the number of protons, they arrange the elements in order of their increasing atomic number (Z). The modern periodic table has seven horizontal rows (called periods) and 18 vertical columns (called groups). The properties of the elements in a particular row vary across it, providing the concept of periodicity.

There are several versions of the periodic table used in modern science. The International Union of Pure and Applied Chemistry (IUPAC)

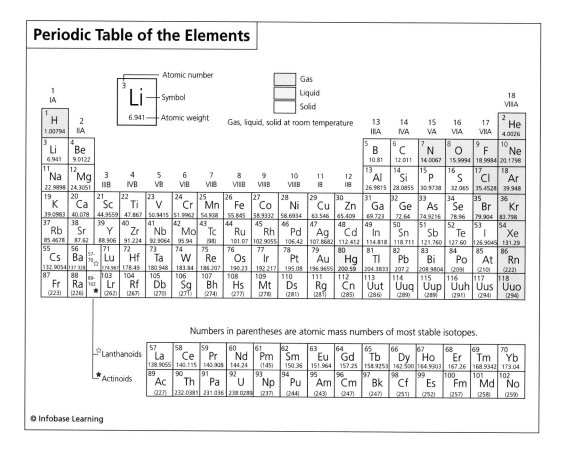

recommends labeling the vertical columns from 1 to 18, starting with hydrogen (H) as the top of group 1 and ending with helium (He) as the top of group 18. The IUPAC further recommends labeling the periods (rows) from 1 to 7. Hydrogen (H) and helium (He) are the only two elements found in period (row) 1. Period 7 starts with francium (Fr) and includes the actinide series as well as the transactinides (very short-lived, human-made, super-heavy elements).

The row (or period) in which an element appears in the periodic table tells scientists how many electron shells an atom of that particular element possesses. The column (or group) lets scientists know how many electrons to expect in an element's outermost electron shell. Scientists call an electron residing in an atom's outermost shell a valence electron. Chemists have learned that it is these valence electrons that determine the chemistry of a particular element. The periodic table is structured such that all the elements in the same column (group) have the same number of valence electrons. The elements that appear in a particular column (group) display similar chemistry.

ELEMENTS LISTED BY ATOMIC NUMBER

1	H	Hydrogen	14	Si	Silicon
2	He	Helium	15	P	Phosphorus
3	Li	Lithium	16	S	Sulfur
4	Be	Beryllium	17	Cl	Chlorine
5	B	Boron	18	Ar	Argon
6	C	Carbon	19	K	Potassium
7	N	Nitrogen	20	Ca	Calcium
8	O	Oxygen	21	Sc	Scandium
9	F	Fluorine	22	Ti	Titanium
10	Ne	Neon	23	V	Vanadium
11	Na	Sodium	24	Cr	Chromium
12	Mg	Magnesium	25	Mn	Manganese
13	Al	Aluminum	26	Fe	Iron

(continues)

ELEMENTS LISTED BY ATOMIC NUMBER (continued)

27	Co	Cobalt		56	Ba	Barium
28	Ni	Nickel		57	La	Lanthanum
29	Cu	Copper		58	Ce	Cerium
30	Zn	Zinc		59	Pr	Praseodymium
31	Ga	Gallium		60	Nd	Neodymium
32	Ge	Germanium		61	Pm	Promethium
33	As	Arsenic		62	Sm	Samarium
34	Se	Selenium		63	Eu	Europium
35	Br	Bromine		64	Gd	Gadolinium
36	Kr	Krypton		65	Tb	Terbium
37	Rb	Rubidium		66	Dy	Dysprosium
38	Sr	Strontium		67	Ho	Holmium
39	Y	Yttrium		68	Er	Erbium
40	Zr	Zirconium		69	Tm	Thulium
41	Nb	Niobium		70	Yb	Ytterbium
42	Mo	Molybdenum		71	Lu	Lutetium
43	Tc	Technetium		72	Hf	Hafnium
44	Ru	Ruthenium		73	Ta	Tantalum
45	Rh	Rhodium		74	W	Tungsten
46	Pd	Palladium		75	Re	Rhenium
47	Ag	Silver		76	Os	Osmium
48	Cd	Cadmium		77	Ir	Iridium
49	In	Indium		78	Pt	Platinum
50	Sn	Tin		79	Au	Gold
51	Sb	Antimony		80	Hg	Mercury
52	Te	Tellurium		81	Tl	Thallium
53	I	Iodine		82	Pb	Lead
54	Xe	Xenon		83	Bi	Bismuth
55	Cs	Cesium		84	Po	Polonium

85	At	Astatine	102	No	Nobelium
86	Rn	Radon	103	Lr	Lawrencium
87	Fr	Francium	104	Rf	Rutherfordium
88	Ra	Radium	105	Db	Dubnium
89	Ac	Actinium	106	Sg	Seaborgium
90	Th	Thorium	107	Bh	Bohrium
91	Pa	Protactinium	108	Hs	Hassium
92	U	Uranium	109	Mt	Meitnerium
93	Np	Neptunium	110	Ds	Darmstadtium
94	Pu	Plutonium	111	Rg	Roentgenium
95	Am	Americium	112	Cn	Copernicum
96	Cm	Curium	113	Uut	Ununtrium
97	Bk	Berkelium	114	Uuq	Ununquadium
98	Cf	Californium	115	Uup	Ununpentium
99	Es	Einsteinium	116	Uuh	Ununhexium
100	Fm	Fermium	117	Uus	Ununseptium
101	Md	Mendelevium	118	Uuo	Ununoctium

Chronology

Civilization is essentially the story of the human mind understanding and gaining control over matter. The chronology presents some of the major milestones, scientific breakthroughs, and technical developments that formed the modern understanding of matter. Note that dates prior to 1543 are approximate.

13.7 BILLION YEARS AGO..... Big bang event starts the universe.

13.3 BILLION YEARS AGO..... The first stars form and begin to shine intensely.

4.5 BILLION YEARS AGO...... Earth forms within the primordial solar nebula.

3.6 BILLION YEARS AGO...... Life (simple microorganisms) appears in Earth's oceans.

2,000,000–100,000 B.C.E... Early hunters of the Lower Paleolithic learn to use simple stone tools, such as handheld axes.

100,000–40,000 B.C.E..... Neanderthal man of Middle Paleolithic lives in caves, controls fire, and uses improved stone tools for hunting.

40,000–10,000 B.C.E...... During the Upper Paleolithic, Cro-Magnon man displaces Neanderthal man. Cro-Magnon people develop more organized hunting and fishing activities using improved stone tools and weapons.

8000–3500 B.C.E.......... Neolithic Revolution takes place in the ancient Middle East as people shift their dependence for subsistence from hunting and gathering to crop cultivation and animal domestication.

3500–1200 B.C.E.......... Bronze Age occurs in the ancient Middle East, when metalworking artisans start using bronze (a copper and tin alloy) to make weapons and tools.

1200–600 B.C.E........... People in the ancient Middle East enter the Iron Age. Eventually, the best weapons and tools are made of steel, an alloy of iron and varying amounts of carbon. The improved metal tools and weapons spread to Greece and later to Rome.

1000 B.C.E. .... By this time, people in various ancient civilizations have discovered and are using the following chemical elements (in alphabetical order): carbon (C), copper (Cu), gold (Au), iron (Fe), lead (Pb), mercury (Hg), silver (Ag), sulfur (S), tin (Sn), and zinc (Zn).

650 B.C.E. .... Kingdom of Lydia introduces officially minted gold and silver coins.

600 B.C.E. .... Early Greek philosopher Thales of Miletus postulates that all substances come from water and would eventually turn back into water.

450 B.C.E. .... Greek philosopher Empedocles proposes that all matter is made up of four basic elements (earth, air, water, and fire) that periodically combine and separate under the influence of two opposing forces (love and strife).

430 B.C.E. .... Greek philosopher Democritus proposes that all things consist of changeless, indivisible, tiny pieces of matter called *atoms*.

250 B.C.E. .... Archimedes of Syracuse designs an endless screw, later called the Archimedes screw. People use the fluid-moving device to remove water from the holds of sailing ships and to irrigate arid fields.

300 C.E. .... Greek alchemist Zosimos of Panoplis writes the oldest known work describing alchemy.

850 .... The Chinese use gunpowder for festive fireworks. It is a mixture of sulfur (S), charcoal (C), and potassium nitrate (KNO_3).

1247 .... British monk Roger Bacon writes the formula for gunpowder in his encyclopedic work *Opus Majus*.

1250 .... German theologian and natural philosopher Albertus Magnus isolates the element arsenic (As).

1439 .... Johannes Gutenberg successfully incorporates movable metal type in his mechanical printing press.

His revolutionary approach to printing depends on a durable, hard metal alloy called type metal, which consists of a mixture of lead (Pb), tin (Sn), and antimony (Sb).

1543 .... Start of the Scientific Revolution. Polish astronomer Nicholas Copernicus promotes heliocentric (Sun-centered) cosmology with his deathbed publication of *On the Revolutions of Celestial Orbs.*

1638 .... Italian scientist Galileo Galilei publishes extensive work on solid mechanics, including uniform acceleration, free fall, and projectile motion.

1643 .... Italian physicist Evangelista Torricelli designs the first mercury barometer and then records the daily variation of atmospheric pressure.

1661 .... Irish-British scientist Robert Boyle publishes *The Sceptical Chymist,* in which he abandons the four classical Greek elements (earth, air, water, and fire) and questions how alchemists determine what substances are elements.

1665 .... British scientist Robert Hooke publishes *Micrographia,* in which he describes pioneering applications of the optical microscope in chemistry, botany, and other scientific fields.

1667 .... The work of German alchemist Johann Joachim Becher forms the basis of the phlogiston theory of heat.

1669 .... German alchemist Hennig Brand discovers the element phosphorous (P).

1678 .... Robert Hooke studies the action of springs and reports that the extension (or compression) of an elastic material takes place in direct proportion to the force exerted on the material.

1687 .... British physicist Sir Isaac Newton publishes *The Principia.* His work provides the mathematical foundations for understanding (from a classical physics perspective) the motion of almost everything in the physical universe.

1738 .... Swiss mathematician Daniel Bernoulli publishes *Hydrodynamica.* In this seminal work, he identifies the relationships between density, pressure, and velocity in flowing fluids.

1748 .... While conducting experiments with electricity, American statesman and scientist Benjamin Franklin coins the term *battery.*

1754 .... Scottish chemist Joseph Black discovers a new gaseous substance, which he calls "fixed air." Other scientists later identify it as carbon dioxide (CO_2).

1764 .... Scottish engineer James Watt greatly improves the Newcomen steam engine. Watt steam engines power the First Industrial Revolution.

1772 .... Scottish physician and chemist Daniel Rutherford isolates a new colorless gaseous substance, calling it "noxious air." Other scientists soon refer to the new gas as nitrogen (N_2).

1785 .... French scientist Charles-Augustin de Coulomb performs experiments that lead to the important law of electrostatics, later known as Coulomb's law.

1789 .... French chemist Antoine-Laurent Lavoisier publishes *Treatise of Elementary Chemistry,* the first modern textbook on chemistry. Lavoisier also promotes the caloric theory of heat.

1800 .... Italian physicist Count Alessandro Volta invents the voltaic pile. His device is the forerunner of the modern electric battery.

1803 .... British schoolteacher and chemist John Dalton revives the atomic theory of matter. From his experiments, he concludes that all matter consists of combinations of atoms and that all the atoms of a particular element are identical.

1807 .... British chemist Sir Humphry Davy discovers the element potassium (K) while experimenting with caustic potash (KOH). Potassium is the first metal isolated by the process of electrolysis.

1811 . . . . Italian physicist Amedeo Avogadro proposes that equal volumes of different gases under the same conditions of pressure and temperature contain the same number of molecules. Scientists call this important hypothesis Avogadro's law.

1820 . . . . Danish physicist Hans Christian Ørsted discovers a relationship between magnetism and electricity.

1824 . . . . French military engineer Sadi Carnot publishes *Reflections on the Motive Power of Fire.* Despite the use of caloric theory, his work correctly identifies the general thermodynamic principles that govern the operation and efficiency of all heat engines.

1826 . . . . French scientist André-Marie Ampère experimentally formulates the relationship between electricity and magnetism.

1827 . . . . Experiments performed by German physicist George Simon Ohm indicate a fundamental relationship among voltage, current, and resistance.

1828 . . . . Swedish chemist Jöns Jacob Berzelius discovers the element thorium (Th).

1831 . . . . British experimental scientist Michael Faraday discovers the principle of electromagnetic induction. This principle is the basis for the electric dynamo.

Independent of Faraday, the American physicist Joseph Henry publishes a paper describing the electric motor (essentially a reverse dynamo).

1841 . . . . German physicist and physician Julius Robert von Mayer states the conservation of energy principle, namely that energy can neither be created nor destroyed.

1847 . . . . British physicist James Prescott Joule experimentally determines the mechanical equivalent of heat.

Joule's work is a major step in developing the modern science of thermodynamics.

1866 .... Swedish scientist-industrialist Alfred Nobel finds a way to stabilize nitroglycerin and calls the new chemical explosive mixture dynamite.

1869 .... Russian chemist Dmitri Mendeleev introduces a periodic listing of the 63 known chemical elements in *Principles of Chemistry*. His periodic table includes gaps for elements predicted but not yet discovered.

American printer John W. Hyatt formulates celluloid, a flammable thermoplastic material made from a mixture of cellulose nitrate, alcohol, and camphor.

1873 .... Scottish mathematician and theoretical physicist James Clerk Maxwell publishes *Treatise on Electricity and Magnetism.*

1876 .... American physicist and chemist Josiah Willard Gibbs publishes *On the Equilibrium of Heterogeneous Substances*. This compendium forms the theoretical foundation of physical chemistry.

1884 .... Swedish chemist Svante Arrhenius proposes that electrolytes split or dissociate into electrically opposite positive and negative ions.

1888 .... German physicist Heinrich Rudolf Hertz produces and detects radio waves.

1895 .... German physicist Wilhelm Conrad Roentgen discovers X-rays.

1896 .... While investigating the properties of uranium salt, French physicist Antoine-Henri Becquerel discovers radioactivity.

1897 .... British physicist Sir Joseph John Thomson performs experiments that demonstrate the existence of the electron—the first subatomic particle discovered.

1898 .... French scientists Pierre and (Polish-born) Marie Curie announce the discovery of two new radioactive elements, polonium (Po) and radium (Ra).

1900 .... German physicist Max Planck postulates that blackbodies radiate energy only in discrete packets (or quanta) rather than continuously. His hypothesis marks the birth of quantum theory.

1903 .... New Zealand–born British physicist Baron (Ernest) Rutherford and British radiochemist Frederick Soddy propose the law of radioactive decay.

1904 .... German physicist Ludwig Prandtl revolutionizes fluid mechanics by introducing the concept of the boundary layer and its role in fluid flow.

1905 .... Swiss-German-American physicist Albert Einstein publishes the special theory of relativity, including the famous mass-energy equivalence formula ($E = mc^2$).

1907 .... Belgian-American chemist Leo Baekeland formulates Bakelite. This synthetic thermoplastic material ushers in the age of plastics.

1911. .... Baron Ernest Rutherford proposes the concept of the atomic nucleus based on the startling results of an alpha particle–gold foil scattering experiment.

1912 .... German physicist Max von Laue discovers that X-rays are diffracted by crystals.

1913 .... Danish physicist Niels Bohr presents his theoretical model of the hydrogen atom—a brilliant combination of atomic theory with quantum physics.

Frederick Soddy proposes the existence of isotopes.

1914 .... British physicist Henry Moseley measures the characteristic X-ray lines of many chemical elements.

1915 .... Albert Einstein presents his general theory of relativity, which relates gravity to the curvature of space-time.

1919 .... Ernest Rutherford bombards nitrogen (N) nuclei with alpha particles, causing the nitrogen nuclei to transform into oxygen (O) nuclei and to emit protons (hydrogen nuclei).

British physicist Francis Aston uses the newly invented mass spectrograph to identify more than 200 naturally occurring isotopes.

1923 . . . . American physicist Arthur Holly Compton conducts experiments involving X-ray scattering that demonstrate the particle nature of energetic photons.

1924 . . . . French physicist Louis-Victor de Broglie proposes the particle-wave duality of matter.

1926 . . . . Austrian physicist Erwin Schrödinger develops quantum wave mechanics to describe the dual wave-particle nature of matter.

1927 . . . . German physicist Werner Heisenberg introduces his uncertainty principle.

1929 . . . . American astronomer Edwin Hubble announces that his observations of distant galaxies suggest an expanding universe.

1932 . . . . British physicist Sir James Chadwick discovers the neutron.

British physicist Sir John Cockcroft and Irish physicist Ernest Walton use a linear accelerator to bombard lithium (Li) with energetic protons, producing the first artificial disintegration of an atomic nucleus.

American physicist Carl D. Anderson discovers the positron.

1934 . . . . Italian-American physicist Enrico Fermi proposes a theory of beta decay that includes the neutrino. He also starts to bombard uranium with neutrons and discovers the phenomenon of slow neutrons.

1938 . . . . German chemists Otto Hahn and Fritz Strassmann bombard uranium with neutrons and detect the presence of lighter elements. Austrian physicist Lise Meitner and Austrian-British physicist Otto Frisch review Hahn's work and conclude in early 1939 that the German chemists had split the atomic nucleus, achieving neutron-induced nuclear fission.

E.I. du Pont de Nemours & Company introduces a new thermoplastic material called nylon.

1941 .... American nuclear scientist Glenn T. Seaborg and his associates use the cyclotron at the University of California, Berkeley, to synthesize plutonium (Pu).

1942 .... Modern nuclear age begins when Enrico Fermi's scientific team at the University of Chicago achieves the first self-sustained, neutron-induced fission chain reaction at Chicago Pile One (CP-1), a uranium-fueled, graphite-moderated atomic pile (reactor).

1945 .... American scientists successfully detonate the world's first nuclear explosion, a plutonium-implosion device code-named Trinity.

1947 .... American physicists John Bardeen, Walter Brattain, and William Shockley invent the transistor.

1952 .... A consortium of 11 founding countries establishes CERN, the European Organization for Nuclear Research, at a site near Geneva, Switzerland.

United States tests the world's first thermonuclear device (hydrogen bomb) at the Enewetak Atoll in the Pacific Ocean. Code-named Ivy Mike, the experimental device produces a yield of 10.4 megatons.

1964 .... German-American physicist Arno Allen Penzias and American physicist Robert Woodrow Wilson detect the cosmic microwave background (CMB).

1967 .... German-American physicist Hans Albrecht Bethe receives the 1967 Nobel Prize in physics for his theory of thermonuclear reactions being responsible for energy generation in stars.

1969 .... On July 20, American astronauts Neil Armstrong and Edwin "Buzz" Aldrin successfully land on the Moon as part of NASA's *Apollo 11* mission.

1972 .... NASA launches the *Pioneer 10* spacecraft. It eventually becomes the first human-made object to leave the solar system on an interstellar trajectory

1985 .... American chemists Robert F. Curl, Jr., and Richard E. Smalley, collaborating with British astronomer Sir Harold W. Kroto, discover the buckyball, an allotrope of pure carbon.

1996 .... Scientists at CERN (near Geneva, Switzerland) announce the creation of antihydrogen, the first human-made antimatter atom.

1998 .... Astrophysicists investigating very distant Type 1A supernovae discover that the universe is expanding at an accelerated rate. Scientists coin the term *dark energy* in their efforts to explain what these observations physically imply.

2001 .... American physicist Eric A. Cornell, German physicist Wolfgang Ketterle, and American physicist Carl E. Wieman share the 2001 Nobel Prize in physics for their fundamental studies of the properties of Bose-Einstein condensates.

2005 .... Scientists at the Lawrence Livermore National Laboratory (LLNL) in California and the Joint Institute for Nuclear Research (JINR) in Dubna, Russia, perform collaborative experiments that establish the existence of super-heavy element 118, provisionally called ununoctium (Uuo).

2008 .... An international team of scientists inaugurates the world's most powerful particle accelerator, the Large Hadron Collider (LHC), located at the CERN laboratory near Geneva, Switzerland.

2009 .... British scientist Charles Kao, American scientist Willard Boyle, and American scientist George Smith share the 2009 Nobel Prize in physics for their pioneering efforts in fiber optics and imaging semiconductor devices, developments that unleashed the information technology revolution.

2010 . . . . Element 112 is officially named Copernicum (Cn) by the IUPAC in honor of Polish astronomer Nicholas Copernicus (1473–1543), who championed heliocentric cosmology.

Scientists at the Joint Institute for Nuclear Research in Dubna, Russia, announce the synthesis of element 117 (ununseptium [Uus]) in early April.

Glossary

absolute zero   the lowest possible temperature; equal to 0 kelvin (K) (−459.67°F, −273.15°C)

acceleration (a)   rate at which the velocity of an object changes with time

accelerator   device for increasing the velocity and energy of charged elementary particles

acid   substance that produces hydrogen ions (H^+) when dissolved in water

actinoid (formerly actinide)   series of heavy metallic elements beginning with element 89 (actinium) and continuing through element 103 (lawrencium)

activity   measure of the rate at which a material emits nuclear radiations

air   overall mixture of gases that make up Earth's atmosphere

alchemy   mystical blend of sorcery, religion, and prescientific chemistry practiced in many early societies around the world

alloy   solid solution (compound) or homogeneous mixture of two or more elements, at least one of which is an elemental metal

alpha particle (α)   positively charged nuclear particle emitted from the nucleus of certain radioisotopes when they undergo decay; consists of two protons and two neutrons bound together

alternating current (AC)   electric current that changes direction periodically in a circuit

American customary system of units (also American system)   used primarily in the United States; based on the foot (ft), pound-mass (lbm), pound-force (lbf), and second (s). Peculiar to this system is the artificial construct (based on Newton's second law) that one pound-force equals one pound-mass (lbm) at sea level on Earth

ampere (A)   SI unit of electric current

anode   positive electrode in a battery, fuel cell, or electrolytic cell; oxidation occurs at anode

antimatter   matter in which the ordinary nuclear particles are replaced by corresponding antiparticles

Archimedes principle   the fluid mechanics rule that states that the buoyant (upward) force exerted on a solid object immersed in a fluid equals the weight of the fluid displaced by the object

atom   smallest part of an element, indivisible by chemical means; consists of a dense inner core (nucleus) that contains protons and neutrons and a cloud of orbiting electrons

atomic mass   *See* **relative atomic mass**

atomic mass unit (amu)   1/12 mass of carbon's most abundant isotope, namely carbon-12

atomic number (Z)   total number of protons in the nucleus of an atom and its positive charge

atomic weight   the mass of an atom relative to other atoms. *See also* **relative atomic mass**

battery   electrochemical energy storage device that serves as a source of direct current or voltage

becquerel (Bq)   SI unit of radioactivity; one disintegration (or spontaneous nuclear transformation) per second. *Compare with* **curie**

beta particle (β)   elementary particle emitted from the nucleus during radioactive decay; a negatively charged beta particle is identical to an electron

big bang   theory in cosmology concerning the origin of the universe; postulates that about 13.7 billion years ago, an initial singularity experienced a very large explosion that started space and time. Astrophysical observations support this theory and suggest that the universe has been expanding at different rates under the influence of gravity, dark matter, and dark energy

blackbody   perfect emitter and perfect absorber of electromagnetic radiation; radiant energy emitted by a blackbody is a function only of the emitting object's absolute temperature

black hole   incredibly compact, gravitationally collapsed mass from which nothing can escape

boiling point   temperature (at a specified pressure) at which a liquid experiences a change of state into a gas

Bose-Einstein condensate (BEC)   state of matter in which extremely cold atoms attain the same quantum state and behave essentially as a large "super atom"

boson   general name given to any particle with a spin of an integral number (0, 1, 2, etc.) of quantum units of angular momentum. Carrier particles of all interactions are bosons. *See also* **carrier particle**

brass   alloy of copper (Cu) and zinc (Zn)

British thermal unit (Btu)   amount of heat needed to raise the temperature of 1 lbm of water 1°F at normal atmospheric pressure; 1 Btu = 1,055 J = 252 cal

bronze   alloy of copper (Cu) and tin (Sn)

calorie (cal)   quantity of heat; defined as the amount needed to raise one gram of water 1°C at normal atmospheric pressure; 1 cal = 4.1868 J = 0.004 Btu

carbon dioxide (CO_2)   colorless, odorless, noncombustible gas present in Earth's atmosphere

Carnot cycle   ideal reversible thermodynamic cycle for a theoretical heat engine; represents the best possible thermal efficiency of any heat engine operating between two absolute temperatures (T_1 and T_2)

carrier particle   within the standard model, gluons are carrier particles for strong interactions; photons are carrier particles of electromagnetic interactions; and the W and Z bosons are carrier particles for weak interactions. *See also* **standard model**

catalyst   substance that changes the rate of a chemical reaction without being consumed or changed by the reaction

cathode   negative electrode in a battery, fuel cell, electrolytic cell, or electron (discharge) tube through which a primary stream of electrons enters a system

chain reaction   reaction that stimulates its own repetition. *See also* **nuclear chain reaction**

change of state   the change of a substance from one physical state to another; the atoms or molecules are structurally rearranged without experiencing a change in composition. Sometimes called change of phase or phase transition

charged particle   elementary particle that carries a positive or negative electric charge

chemical bond(s)   force(s) that holds atoms together to form stable configurations of molecules

chemical property   characteristic of a substance that describes the manner in which the substance will undergo a reaction with another substance, resulting in a change in chemical composition. *Compare with* **physical property**

chemical reaction   involves changes in the electron structure surrounding the nucleus of an atom; a dissociation, recombination, or rearrangement of atoms. During a chemical reaction, one or more kinds of matter (called reactants) are transformed into one or several new kinds of matter (called products)

color charge   in the standard model, the charge associated with strong interactions. Quarks and gluons have color charge and thus participate in strong interactions. Leptons, photons, W bosons, and Z bosons do not have color charge and consequently do not participate in strong interactions. *See also* **standard model**

combustion   chemical reaction (burning or rapid oxidation) between a fuel and oxygen that generates heat and usually light

composite materials   human-made materials that combine desirable properties of several materials to achieve an improved substance; includes combinations of metals, ceramics, and plastics with built-in strengthening agents

compound   pure substance made up of two or more elements chemically combined in fixed proportions

compressible flow   fluid flow in which density changes cannot be neglected

compression   condition when an applied external force squeezes the atoms of a material closer together. *Compare with* **tension**

concentration   for a solution, the quantity of dissolved substance per unit quantity of solvent

condensation   change of state process by which a vapor (gas) becomes a liquid. *The opposite of* **evaporation**

conduction (thermal)   transport of heat through an object by means of a temperature difference from a region of higher temperature to a region of lower temperature. *Compare with* **convection**

conservation of mass and energy   Einstein's special relativity principle stating that energy (E) and mass (m) can neither be created nor destroyed, but are interchangeable in accordance with the equation $E = mc^2$, where c represents the speed of light

convection   fundamental form of heat transfer characterized by mass motions within a fluid resulting in the transport and mixing of the properties of that fluid

coulomb (C)   SI unit of electric charge; equivalent to quantity of electric charge transported in one second by a current of one ampere

covalent bond   the chemical bond created within a molecule when two or more atoms share an electron

creep   slow, continuous, permanent deformation of solid material caused by a constant tensile or compressive load that is less than the load necessary for the material to give way (yield) under pressure. *See also* **plastic deformation**

crystal   a solid whose atoms are arranged in an orderly manner, forming a distinct, repetitive pattern

curie (Ci)   traditional unit of radioactivity equal to 37 billion (37×10^9) disintegrations per second. *Compare with* **becquerel**

current (I)   flow of electric charge through a conductor

dark energy   a mysterious natural phenomenon or unknown cosmic force thought responsible for the observed acceleration in the rate of expansion of the universe. Astronomical observations suggest dark energy makes up about 72 percent of the universe

dark matter (nonbaryonic matter)   exotic form of matter that emits very little or no electromagnetic radiation. It experiences no measurable interaction with ordinary (baryonic) matter but somehow accounts for the observed structure of the universe. It makes up about 23 percent of the content of the universe, while ordinary matter makes up less than 5 percent

density (ρ)   mass of a substance per unit volume at a specified temperature

deposition   direct transition of a material from the gaseous (vapor) state to the solid state without passing through the liquid phase. *Compare with* **sublimation**

dipole magnet   any magnet with one north and one south pole

direct current (DC)   electric current that always flows in the same direction through a circuit

elastic deformation   temporary change in size or shape of a solid due to an applied force (stress); when force is removed the solid returns to its original size and shape

elasticity   ability of a body that has been deformed by an applied force to return to its original shape when the force is removed

elastic modulus   a measure of the stiffness of a solid material; defined as the ratio of stress to strain

electricity   flow of energy due to the motion of electric charges; any physical effect that results from the existence of moving or stationary electric charges

electrode   conductor (terminal) at which electricity passes from one medium into another; positive electrode is the *anode;* negative electrode is the *cathode*

electrolyte   a chemical compound that, in an aqueous (water) solution, conducts an electric current

electromagnetic radiation (EMR)   oscillating electric and magnetic fields that propagate at the speed of light. Includes in order of increasing frequency and energy: radio waves, radar waves, infrared (IR) radiation, visible light, ultraviolet radiation, X-rays, and gamma rays

electron (e)   stable elementary particle with a unit negative electric charge (1.602×10^{-19} C). Electrons form an orbiting cloud, or shell, around the positively charged atomic nucleus and determine an atom's chemical properties

electron volt (eV)   energy gained by an electron as it passes through a potential difference of one volt; one electron volt has an energy equivalence of 1.519×10^{-22} Btu $= 1.602 \times 10^{-19}$ J

element   pure chemical substance indivisible into simpler substances by chemical means; all the atoms of an element have the same number of protons in the nucleus and the same number of orbiting electrons, although the number of neutrons in the nucleus may vary

elementary particle   a fundamental constituent of matter; the basic atomic model suggests three elementary particles: the proton, neutron, and electron. *See also* **fundamental particle**

endothermic reaction   chemical reaction requiring an input of energy to take place. *Compare with* **exothermic reaction**

energy (E)   capacity to do work; appears in many different forms, such as mechanical, thermal, electrical, chemical, and nuclear

entropy (S)   measure of disorder within a system; as entropy increases, energy becomes less available to perform useful work

evaporation   physical process by which a liquid is transformed into a gas (vapor) at a temperature below the boiling point of the liquid. *Compare with* **sublimation**

excited state   state of a molecule, atom, electron, or nucleus when it possesses more than its normal energy. *Compare with* **ground state**

exothermic reaction chemical reaction that releases energy as it takes place. *Compare with* **endothermic reaction**

fatigue weakening or deterioration of metal or other material that occurs under load, especially under repeated cyclic or continued loading

fermion general name scientists give to a particle that is a matter constituent. Fermions are characterized by spin in odd half-integer quantum units (namely, 1/2, 3/2, 5/2, etc.); quarks, leptons, and baryons are all fermions

fission (nuclear) splitting of the nucleus of a heavy atom into two lighter nuclei accompanied by the release of a large amount of energy as well as neutrons, X-rays, and gamma rays

flavor in the standard model, quantum number that distinguishes different types of quarks and leptons. *See also* **quark; lepton**

fluid mechanics scientific discipline that deals with the behavior of fluids (both gases and liquids) at rest (fluid statics) and in motion (fluid dynamics)

foot-pound (force) (ft-lb$_{force}$) unit of work in American customary system of units; 1 ft-lb$_{force}$ = 1.3558 J

force (F) the cause of the acceleration of material objects as measured by the rate of change of momentum produced on a free body. Force is a vector quantity mathematically expressed by Newton's second law of motion: force = mass × acceleration

freezing point the temperature at which a substance experiences a change from the liquid state to the solid state at a specified pressure; at this temperature, the solid and liquid states of a substance can coexist in equilibrium. *Synonymous with* **melting point**

fundamental particle particle with no internal substructure; in the standard model, any of the six types of quarks or six types of leptons and their antiparticles. Scientists postulate that all other particles are made from a combination of quarks and leptons. *See also* **elementary particle**

fusion (nuclear) nuclear reaction in which lighter atomic nuclei join together (fuse) to form a heavier nucleus, liberating a great deal of energy

g acceleration due to gravity at sea level on Earth; approximately 32.2 ft/s^2 (9.8 m/s^2)

gamma ray (γ)   high-energy, very short–wavelength photon of electromagnetic radiation emitted by a nucleus during certain nuclear reactions or radioactive decay

gas   state of matter characterized as an easily compressible fluid that has neither a constant volume nor a fixed shape; a gas assumes the total size and shape of its container

gravitational lensing   bending of light from a distant celestial object by a massive (gravitationally influential) foreground object

ground state   state of a nucleus, atom, or molecule at its lowest (normal) energy level

hadron   any particle (such as a baryon) that exists within the nucleus of an atom; made up of quarks and gluons, hadrons interact with the strong force

half-life (radiological)   time in which half the atoms of a particular radioactive isotope disintegrate to another nuclear form

heat   energy transferred by a temperature difference or thermal process. *Compare* **work**

heat capacity (c)   amount of heat needed to raise the temperature of an object by one degree

heat engine   thermodynamic system that receives energy in the form of heat and that, in the performance of energy transformation on a working fluid, does work. Heat engines function in thermodynamic cycles

hertz (Hz)   SI unit of frequency; equal to one cycle per second

high explosive (HE)   energetic material that detonates (rather than burns); the rate of advance of the reaction zone into the unreacted material exceeds the velocity of sound in the unreacted material

horsepower (hp)   American customary system unit of power; 1 hp = 550 ft-lb$_{force}$/s = 746 W

hydraulic   operated, moved, or affected by liquid used to transmit energy

hydrocarbon   organic compound composed of only carbon and hydrogen atoms

ideal fluid   *See* **perfect fluid**

ideal gas law   important physical principle: $P V = n R_u T$, where P is pressure, V is volume, T is temperature, n is the number of moles of gas, and R_u is the universal gas constant

incompressible flow fluid flow in which density changes can be neglected. *Compare with* **compressible flow**

inertia resistance of a body to a change in its state of motion

infrared (IR) radiation that portion of the electromagnetic (EM) spectrum lying between the optical (visible) and radio wavelengths

International System of units *See* **SI unit system**

inviscid fluid perfect fluid that has zero coefficient of viscosity

ion atom or molecule that has lost or gained one or more electrons, so that the total number of electrons does not equal the number of protons

ionic bond formed when one atom gives up at least one outer electron to another atom, creating a chemical bond–producing electrical attraction between the atoms

isotope atoms of the same chemical element but with different numbers of neutrons in their nucleus

joule (J) basic unit of energy or work in the SI unit system; 1 J = 0.2388 calorie = 0.00095 Btu

kelvin (K) SI unit of absolute thermodynamic temperature

kinetic energy (KE) energy due to motion

lepton fundamental particle of matter that does not participate in strong interactions; in the standard model, the three charged leptons are the electron (e), the muon (μ), and the tau (τ) particle; the three neutral leptons are the electron neutrino (v_e), the muon neutrino (v_μ), and the tau neutrino (v_τ). A corresponding set of antiparticles also exists. *See also* **standard model**

light-year (ly) distance light travels in one year; 1 ly $\approx$ 5.88 $\times$ 10^{12} miles (9.46 $\times$ 10^{12} km)

liquid state of matter characterized as a relatively incompressible flowing fluid that maintains an essentially constant volume but assumes the shape of its container

liter (l or L) SI unit of volume; 1 L = 0.264 gal

magnet material or device that exhibits magnetic properties capable of causing the attraction or repulsion of another magnet or the attraction of certain ferromagnetic materials such as iron

manufacturing process of transforming raw material(s) into a finished product, especially in large quantities

mass (m)   property that describes how much material makes up an object and gives rise to an object's inertia

mass number   *See* **relative atomic mass**

mass spectrometer   instrument that measures relative atomic masses and relative abundances of isotopes

material   tangible substance (chemical, biological, or mixed) that goes into the makeup of a physical object

mechanics   branch of physics that deals with the motions of objects

melting point   temperature at which a substance experiences a change from the solid state to the liquid state at a specified pressure; at this temperature, the solid and liquid states of a substance can coexist in equilibrium. *Synonymous with* **freezing point**

metallic bond   chemical bond created as many atoms of a metallic substance share the same electrons

meter (m)   fundamental SI unit of length; 1 meter = 3.281 feet. British spelling *metre*

metric system   *See* **SI unit system**

metrology   science of dimensional measurement; sometimes includes the science of weighing

microwave (radiation)   comparatively short-wavelength electromagnetic (EM) wave in the radio frequency portion of the EM spectrum

mirror matter   *See* **antimatter**

mixture   a combination of two or more substances, each of which retains its own chemical identity

molarity (M)   concentration of a solution expressed as moles of solute per kilogram of solvent

mole (mol)   SI unit of the amount of a substance; defined as the amount of substance that contains as many elementary units as there are atoms in 0.012 kilograms of carbon-12, a quantity known as Avogadro's number (N_A), which has a value of about 6.022×10^{23} molecules/mole

molecule   smallest amount of a substance that retains the chemical properties of the substance; held together by chemical bonds, a molecule can consist of identical atoms or different types of atoms

monomer   substance of relatively low molecular mass; any of the small molecules that are linked together by covalent bonds to form a polymer

natural material   material found in nature, such as wood, stone, gases, and clay

neutrino (ν)   lepton with no electric charge and extremely low (if not zero) mass; three known types of neutrinos are the electron neutrino (v_e), the muon neutrino (v_μ), and the tau neutrino (v_τ). *See also* **lepton**

neutron (n)   an uncharged elementary particle found in the nucleus of all atoms except ordinary hydrogen. Within the standard model, the neutron is a baryon with zero electric charge consisting of two down (d) quarks and one up (u) quark. *See also* **standard model**

newton (N)   The SI unit of force; 1 N = 0.2248 lbf

nuclear chain reaction   occurs when a fissionable nuclide (such as plutonium-239) absorbs a neutron, splits (or fissions), and releases several neutrons along with energy. A fission chain reaction is self-sustaining when (on average) at least one released neutron per fission event survives to create another fission reaction

nuclear energy   energy released by a nuclear reaction (fission or fusion) or by radioactive decay

nuclear radiation   particle and electromagnetic radiation emitted from atomic nuclei as a result of various nuclear processes, such as radioactive decay and fission

nuclear reaction   reaction involving a change in an atomic nucleus, such as fission, fusion, neutron capture, or radioactive decay

nuclear reactor   device in which a fission chain reaction can be initiated, maintained, and controlled

nuclear weapon   precisely engineered device that releases nuclear energy in an explosive manner as a result of nuclear reactions involving fission, fusion, or both

nucleon   constituent of an atomic nucleus; a proton or a neutron

nucleus   (plural: nuclei) small, positively charged central region of an atom that contains essentially all of its mass. All nuclei contain both protons and neutrons except the nucleus of ordinary hydrogen, which consists of a single proton

nuclide   general term applicable to all atomic (isotopic) forms of all the elements; nuclides are distinguished by their atomic number, relative mass number (atomic mass), and energy state

ohm (Ω)   SI unit of electrical resistance

oxidation   chemical reaction in which oxygen combines with another substance, and the substance experiences one of three processes: (1) the gaining of oxygen, (2) the loss of hydrogen, or (3) the loss of electrons. In these
reactions, the substance being "oxidized" loses electrons and forms positive ions. *Compare with* **reduction**

oxidation-reduction (redox) reaction   chemical reaction in which electrons are transferred between species or in which atoms change oxidation
number

particle   minute constituent of matter, generally one with a measurable
mass

pascal (Pa)   SI unit of pressure; $1 \text{ Pa} = 1 \text{ N/m}^2 = 0.000145 \text{ psi}$

Pascal's principle   when an enclosed (static) fluid experiences an increase
in pressure, the increase is transmitted throughout the fluid; the physical
principle behind all hydraulic systems

Pauli exclusion principle   postulate that no two electrons in an atom can
occupy the same quantum state at the same time; also applies to protons
and neutrons

perfect fluid   hypothesized fluid primarily characterized by a lack of viscosity and usually by incompressibility

perfect gas law   *See* **ideal gas law**

periodic table   list of all the known elements, arranged in rows (periods)
in order of increasing atomic numbers and columns (groups) by similar
physical and chemical characteristics

phase   one of several different homogeneous materials present in a portion of matter under study; the set of states of a large-scale (macroscopic)
physical system having relatively uniform physical properties and chemical composition

phase transition   *See* **change of state**

photon   A unit (or particle) of electromagnetic radiation that carries a
quantum (packet) of energy that is characteristic of the particular radiation. Photons travel at the speed of light and have an effective momentum,
but no mass or electrical charge. In the standard model, a photon is the
carrier particle of electromagnetic radiation

photovoltaic cell   *See* **solar cell**

physical property   characteristic quality of a substance that can be measured or demonstrated without changing the composition or chemical

identity of the substance, such as temperature and density. *Compare with* **chemical property**

Planck's constant (h)   fundamental physical constant describing the extent to which quantum mechanical behavior influences nature. Equals the ratio of a photon's energy (E) to its frequency (ν), namely: h = E/ν = 6.626 × 10^{-34} J-s (6.282 × 10^{-37} Btu-s). *See also* **uncertainty principle**

plasma   electrically neutral gaseous mixture of positive and negative ions; called the fourth state of matter

plastic deformation   permanent change in size or shape of a solid due to an applied force (stress)

plasticity   tendency of a loaded body to assume a (deformed) state other than its original state when the load is removed

plastics   synthesized family of organic (mainly hydrocarbon) polymer materials used in nearly every aspect of modern life

pneumatic   operated, moved, or effected by a pressurized gas (typically air) that is used to transmit energy

polymer   very large molecule consisting of a number of smaller molecules linked together repeatedly by covalent bonds, thereby forming long chains

positron (e^+ or β^+)   elementary antimatter particle with the mass of an electron but charged positively

pound-force (lbf)   basic unit of force in the American customary system; 1 lbf = 4.448 N

pound-mass (lbm)   basic unit of mass in the American customary system; 1 lbm = 0.4536 kg

power   rate with respect to time at which work is done or energy is transformed or transferred to another location; 1 hp = 550 ft-lb$_{force}$/s = 746 W

pressure (P)   the normal component of force per unit area exerted by a fluid on a boundary; 1 psi = 6,895 Pa

product   substance produced by or resulting from a chemical reaction

proton (p)   stable elementary particle with a single positive charge. In the the standard model, the proton is a baryon with an electric charge of +1; it consists of two up (u) quarks and one down (d) quark. *See also* **standard model**

quantum mechanics   branch of physics that deals with matter and energy on a very small scale; physical quantities are restricted to discrete values and energy to discrete packets called quanta

quark fundamental matter particle that experiences strong-force interactions. The six flavors of quarks in order of increasing mass are up (u), down (d), strange (s), charm (c), bottom (b), and top (t)

radiation heat transfer The transfer of heat by electromagnetic radiation that arises due to the temperature of a body; can takes place in and through a vacuum

radioactive isotope unstable isotope of an element that decays or disintegrates spontaneously, emitting nuclear radiation; also called radioisotope

radioactivity spontaneous decay of an unstable atomic nucleus, usually accompanied by the emission of nuclear radiation, such as alpha particles, beta particles, gamma rays, or neutrons

radio frequency (RF) a frequency at which electromagnetic radiation is useful for communication purposes; specifically, a frequency above 10,000 hertz (Hz) and below 3×10^{11} Hz

rankine (R) American customary unit of absolute temperature. *See also* **kelvin (K)**

reactant original substance or initial material in a chemical reaction

reduction portion of an oxidation-reduction (redox) reaction in which there is a gain of electrons, a gain in hydrogen, or a loss of oxygen. *See also* **oxidation-reduction (redox) reaction**

relative atomic mass (A) total number of protons and neutrons (nucleons) in the nucleus of an atom. Previously called *atomic mass* or *atomic mass number. See also* **atomic mass unit**

residual electromagnetic effect force between electrically neutral atoms that leads to the formation of molecules

residual strong interaction interaction responsible for the nuclear binding force—that is, the strong force holding hadrons (protons and neutrons) together in the atomic nucleus. *See also* **strong force**

resilience property of a material that enables it to return to its original shape and size after deformation

resistance (R) the ratio of the voltage (V) across a conductor to the electric current (I) flowing through it

scientific notation A method of expressing powers of 10 that greatly simplifies writing large numbers; for example, $3 \times 10^6 = 3,000,000$

SI unit system international system of units (the metric system), based upon the meter (m), kilogram (kg), and second (s) as the fundamental units of length, mass, and time, respectively

solar cell (photovoltaic cell)   a semiconductor direct energy conversion device that transforms sunlight into electric energy

solid   state of matter characterized by a three-dimensional regularity of structure; a solid is relatively incompressible, maintains a fixed volume, and has a definitive shape

solution   When scientists dissolve a substance in a pure liquid, they refer to the dissolved substance as the *solute* and the host pure liquid as the *solvent.* They call the resulting intimate mixture the solution

spectroscopy   study of spectral lines from various atoms and molecules; emission spectroscopy infers the material composition of the objects that emitted the light; absorption spectroscopy infers the composition of the intervening medium

speed of light (c)   speed at which electromagnetic radiation moves through a vacuum; regarded as a universal constant equal to 186,283.397 mi/s (299,792.458 km/s)

stable isotope   isotope that does not undergo radioactive decay

standard model   contemporary theory of matter, consisting of 12 fundamental particles (six quarks and six leptons), their respective antiparticles, and four force carriers (gluons, photons, W bosons, and Z bosons)

state of matter   form of matter having physical properties that are quantitatively and qualitatively different from other states of matter; the three more common states on Earth are solid, liquid, and gas

steady state   condition of a physical system in which parameters of importance (fluid velocity, temperature, pressure, etc.) do not vary significantly with time

strain   the change in the shape or dimensions (volume) of an object due to applied forces; longitudinal, volume, and shear are the three basic types of strain

stress   applied force per unit area that causes an object to deform (experience strain); the three basic types of stress are compressive (or tensile) stress, hydrostatic pressure, and shear stress

string theory   theory of quantum gravity that incorporates Einstein's general relativity with quantum mechanics in an effort to explain space-time phenomena on the smallest imaginable scales; vibrations of incredibly tiny stringlike structures form quarks and leptons

strong force   In the standard model, the fundamental force between quarks and gluons that makes them combine to form hadrons, such as protons and neutrons; also holds hadrons together in a nucleus. *See also* **standard model**

subatomic particle   any particle that is small compared to the size of an atom

sublimation   direct transition of a material from the solid state to the gaseous (vapor) state without passing through the liquid phase. *Compare with* **deposition**

superconductivity   the ability of a material to conduct electricity without resistance at a temperature above absolute zero

temperature (T)   thermodynamic property that serves as a macroscopic measure of atomic and molecular motions within a substance; heat naturally flows from regions of higher temperature to regions of lower temperature

tension   condition when applied external forces pull atoms of a material farther apart. *Compare with* **compression**

thermal conductivity (k)   intrinsic physical property of a substance; a material's ability to conduct heat as a consequence of molecular motion

thermodynamics   branch of science that treats the relationships between heat and energy, especially mechanical energy

thermodynamic system   collection of matter and space with boundaries defined in such a way that energy transfer (as work and heat) from and to the system across these boundaries can be easily identified and analyzed

thermometer   instrument or device for measuring temperature

toughness   ability of a material (especially a metal) to absorb energy and deform plastically before fracturing

transmutation   transformation of one chemical element into a different chemical element by a nuclear reaction or series of reactions

transuranic element (isotope)   human-made element (isotope) beyond uranium on the periodic table

ultraviolet (UV) radiation   portion of the electromagnetic spectrum that lies between visible light and X-rays

uncertainty principle   Heisenberg's postulate that places quantum-level limits on how accurately a particle's momentum *(p)* and position *(x)* can be simultaneously measured. Planck's constant (h) expresses this uncertainty as $\Delta x \times \Delta p \geq h/2\pi$

U.S. customary system of units *See* **American customary system of units**

vacuum relative term used to indicate the absence of gas or a region in which there is a very low gas pressure

valence electron electron in the outermost shell of an atom

van der Waals force generally weak interatomic or intermolecular force caused by polarization of electrically neutral atoms or molecules

vapor gaseous state of a substance

velocity vector quantity describing the rate of change of position; expressed as length per unit of time

velocity of light (c) *See* **speed of light**

viscosity measure of the internal friction or flow resistance of a fluid when it is subjected to shear stress

volatile solid or liquid material that easily vaporizes; volatile material has a relatively high vapor pressure at normal temperatures

volt (V) SI unit of electric potential difference

volume (V) space occupied by a solid object or a mass of fluid (liquid or confined gas)

watt (W) SI unit of power (work per unit time); $1 \text{ W} = 1 \text{ J/s} = 0.00134 \text{ hp} = 0.737 \text{ ft-lb}_{force}/\text{s}$

wavelength (λ) the mean distance between two adjacent maxima (or minima) of a wave

weak force fundamental force of nature responsible for various types of radioactive decay

weight (w) the force of gravity on a body; on Earth, product of the mass (m) of a body times the acceleration of gravity (g), namely $w = m \times g$

work (W) energy expended by a force acting though a distance. *Compare with* **heat**

X-ray penetrating form of electromagnetic (EM) radiation that occurs on the EM spectrum between ultraviolet radiation and gamma rays

Further Resources

BOOKS

Allcock, Harry R. *Introduction to Materials Chemistry.* New York: John Wiley & Sons, 2008. A college-level textbook that provides a basic treatment of the principles of chemistry upon which materials science depends.

Angelo, Joseph A., Jr. *Nuclear Technology.* Westport, Conn.: Greenwood Press, 2004. The book provides a detailed discussion of both military and civilian nuclear technology and includes impacts, issues, and future advances.

———. *Encyclopedia of Space and Astronomy.* New York: Facts On File, 2006. Provides a comprehensive treatment of major concepts in astronomy, astrophysics, planetary science, cosmology, and space technology.

Ball, Philip. *Designing the Molecular World: Chemistry at the Frontier.* Princeton, N.J.: Princeton University Press, 1996. Discusses many recent advances in modern chemistry, including nanotechnology and superconductor materials.

———. *Made to Measure: New Materials for the 21st Century.* Princeton, N.J.: Princeton University Press, 1998. Discusses how advanced new materials can significantly influence life in the 21st century.

Bensaude-Vincent, Bernadette, and Isabelle Stengers. *A History of Chemistry.* Cambridge, Mass.: Harvard University Press, 1996. Describes how chemistry emerged as a science and its impact on civilization.

Callister, William D., Jr. *Materials Science and Engineering: An Introduction.* 8th ed. New York: John Wiley & Sons, 2010. Intended primarily for engineers, technically knowledgeable readers will also benefit from this book's introductory treatment of metals, ceramics, polymers, and composite materials.

Charap, John M. *Explaining the Universe: The New Age of Physics.* Princeton, N.J.: Princeton University Press, 2004. Discusses the important discoveries in physics during the 20th century that are influencing civilization.

Close, Frank, et al. *The Particle Odyssey: A Journey to the Heart of the Matter.* New York: Oxford University Press, 2002. A well-illustrated and enjoyable tour of the subatomic world.

Cobb, Cathy, and Harold Goldwhite. *Creations of Fire: Chemistry's Lively History from Alchemy to the Atomic Age.* New York: Plenum Press, 1995. Uses historic circumstances and interesting individuals to describe the emergence of chemistry as a scientific discipline.

Feynman, Richard P. *QED: The Strange Theory of Light and Matter.* Princeton, N.J.: Princeton University Press, 2006. Written by an American Nobel laureate, addresses several key topics in modern physics.

Gordon, J. E. *The New Science of Strong Materials or Why You Don't Fall Through the Floor.* Princeton, N.J.: Princeton University Press, 2006. Discusses the science of structural materials in a manner suitable for both technical and lay audiences.

Hill, John W., and Doris K. Kolb. *Chemistry for Changing Times.* 11th ed. Upper Saddle River, N.J.: Pearson Prentice Hall, 2007. Readable college-level textbook that introduces all the basic areas of modern chemistry.

Krebs, Robert E. *The History and Use of Our Earth's Chemical Elements: A Reference Guide.* 2nd ed. Westport, Conn.: Greenwood Press, 2006. Provides a concise treatment of each of the chemical elements.

Levere, Trevor H. *Transforming Matter: A History of Chemistry from Alchemy to the Buckyball.* Baltimore: Johns Hopkins University Press, 2001. Provides an understandable overview of the chemical sciences from the early alchemists through modern times.

Lutgens, Frederick K., and Edward J. Tarbuck. *The Atmosphere: An Introduction to Meteorology.* 10th ed. Upper Saddle River, N.J.: Pearson Prentice Hall, 2007. Readable college-level textbook that discusses the atmosphere, meteorology, climate, and the physical properties of air.

Mackintosh, Ray, et al. *Nucleus: A Trip into the Heart of Matter.* Baltimore: Johns Hopkins University Press, 2001. Provides a technical though readable explanation of how modern scientists developed their current understanding of the atomic nucleus and the standard model.

Nicolaou, K. C., and Tamsyn Montagnon. *Molecules that Changed the World.* New York: John Wiley & Sons, 2008. Provides an interesting treatment of such important molecules as aspirin, camphor, glucose, quinine, and morphine.

Scerri, Eric R. *The Periodic Table: Its Story and Its Significance.* New York: Oxford University Press, 2007. Provides a detailed look at the periodic table and its iconic role in the practice of modern science.

Smith, William F., and Javad Hashemi. *Foundations of Materials Science and Engineering.* 5th ed. New York: McGraw-Hill, 2006. Provides scientists and engineers of all disciplines an introduction to materials science, including metals, ceramics, polymers, and composite materials. Technically knowledgeable laypersons will find the treatment of specific topics such as biological materials useful.

Strathern, Paul. *Mendeleyev's Dream: The Quest for the Elements.* New York: St. Martin's Press, 2001. Describes the intriguing history of chemistry from the early Greek philosophers to the 19th-century Russian chemist Dmitri Mendeleyev.

Thrower, Peter, and Thomas Mason. *Materials in Today's World.* 3rd ed. New York: McGraw-Hill Companies, 2007. Provides a readable introductory treatment of modern materials science, including biomaterials and nanomaterials.

Trefil, James, and Robert M. Hazen. *Physics Matters: An Introduction to Conceptual Physics.* New York: John Wiley & Sons, 2004. Highly-readable introductory college-level textbook that provides a good overview of physics from classical mechanics to relativity and cosmology. Laypersons will find the treatment of specific topics useful and comprehendible.

Zee, Anthony. *Quantum Field Theory in a Nutshell.* Princeton, N.J.: Princeton University Press, 2003. A reader-friendly treatment of the generally complex and profound physical concepts that constitute quantum field theory.

WEB SITES

To help enrich the content of this book and to make your investigation of matter more enjoyable, the following is a selective list of recommended Web sites. Many of the sites below will also lead to other interesting science-related locations on the Internet. Some sites provide unusual science learning opportunities (such as laboratory simulations) or in-depth educational resources.

American Chemical Society (ACS) is a congressionally chartered independent membership organization that represents professionals at all degree levels and in all fields of science involving chemistry. The ACS Web site includes educational resources for high school and college students. Available online. URL: http://portal.acs.org/portal/acs/corg/content. Accessed on February 12, 2010.

American Institute of Physics (AIP) is a not-for-profit corporation that promotes the advancement and diffusion of the knowledge of physics and its applications to human welfare. This Web site offers an enormous quantity of fascinating information about the history of physics from ancient Greece up to the present day. Available online. URL: http://www.aip.org/aip/. Accessed on February 12, 2010.

Chandra X-ray Observatory (CXO) is a space-based NASA astronomical observatory that observes the universe in the X-ray portion of the electromagnetic spectrum. This Web site contains contemporary information and educational materials about astronomy, astrophysics, and cosmology, including topics such as black holes, neutron stars, dark matter, and dark energy. Available online. URL: http://www.chandra.harvard.edu/. Accessed on February 12, 2010.

The ChemCollective is an online resource for learning about chemistry. Through simulations developed by the Department of Chemistry of Carnegie Mellon University (with funding from the National Science Foundation), a person gets the chance to safely mix chemicals without worrying about accidentally spilling them. Available online. URL: http://www.chemcollective.org/vlab/vlab.php. Accessed on February 12, 2010.

Chemical Heritage Foundation (CHF) maintains a rich and informative collection of materials that describe the history and heritage of the chemical and molecular sciences, technologies, and industries. Available online. URL: http://www.chemheritage.org/. Accessed on February 12, 2010.

Department of Defense (DOD) is responsible for maintaining armed forces of sufficient strength and technology to protect the United States and its citizens from all credible foreign threats. This Web site serves as an efficient access point to activities within the DOD, including those taking place within each of the individual armed services: the U.S. Army, U.S. Navy, U.S. Air Force, and U.S. Marines. As part of national security, the DOD sponsors a large amount of research and development, including activities in materials science, chemistry, physics, and nanotechnology. Available online. URL: http://www.defenselink.mil/. Accessed on February 12, 2010.

Department of Energy (DOE) is the single largest supporter of basic research in the physical sciences in the federal government of the United States. Topics found on this Web site include materials sciences, nanotechnology, energy sciences, chemical science, high-energy physics, and nuclear physics. The Web site also includes convenient links to all of the DOE's national laboratories. Available online. URL: http://energy.gov/. Accessed on February 12, 2010.

Fermi National Accelerator Laboratory (Fermilab) performs research that advances the understanding of the fundamental nature of matter and energy. Fermilab's Web site contains contemporary information about particle physics, the standard model, and the impact of particle physics on society. Available online. URL: http://www.fnal.gov/. Accessed on February 12, 2010.

Hubble Space Telescope (HST) is a space-based NASA observatory that has examined the universe in the (mainly) visible portion of the electromagnetic spectrum. This Web site contains contemporary information and educational materials about astronomy, astrophysics, and cosmology, including topics such as black holes, neutron stars, dark matter, and dark energy. Available online. URL: http://hubblesite.org/. Accessed on February 12, 2010.

Institute and Museum of the History of Science in Florence, Italy, offers a special collection of scientific instruments (some viewable online), including those used by Galileo Galilei. Available online. URL: http://www.imss.fi.it/. Accessed on February 12, 2010.

International Union of Pure and Applied Chemistry (IUPAC) is an international nongovernmental organization that fosters worldwide communications in the chemical sciences and in providing a common language for chemistry that unifies the industrial, academic, and public sectors. Available online. URL: http://www.iupac.org/. Accessed on February 12, 2010.

National Aeronautics and Space Administration (NASA) is the civilian space agency of the U.S. government and was created in 1958 by an act of Congress. NASA's overall mission is to direct, plan, and conduct American civilian (including scientific) aeronautical and space activities for peaceful purposes. Available online. URL: http://www.nasa.gov/. Accessed on February 12, 2010.

National Institute of Standards and Technology (NIST) is an agency of the U.S. Department of Commerce that was founded in 1901 as the nation's first federal physical science research laboratory. The NIST Web site includes contemporary information about many areas of science and engineering, including analytical chemistry, atomic and molecular physics, biometrics, chemical and crystal structure, chemical kinetics, chemistry, construction, environmental data, fire, fluids, material properties, physics, and thermodynamics. Available online. URL: http://www.nist.gov/index.html. Accessed on February 12, 2010.

National Oceanic and Atmospheric Administration (NOAA) was established in 1970 as an agency within the U.S. Department of Commerce to ensure the safety of the general public from atmospheric phenomena and to provide the public with an understanding of Earth's environment and resources. Available online. URL: http://www.noaa.gov/. Accessed on February 12, 2010.

NEWTON: Ask a Scientist is an electronic community for science, math, and computer science educators and students sponsored by the Argonne National Laboratory (ANL) and the U.S. Department of Energy's Office of Science Education. This Web site provides access to a fascinating list of questions and answers involving the following disciplines/topics: astronomy, biology, botany, chemistry, computer science, Earth science, engineering, environmental science, general science, materials science, mathematics, molecular biology, physics, veterinary, weather, and zoology. Available online. URL: http://www.newton.dep.anl.gov/archive.htm. Accessed on February 12, 2010.

Nobel Prizes in Chemistry and Physics. This Web site contains an enormous amount of information about all the Nobel Prizes awarded in physics and chemistry, as well as complementary technical information. Available online. URL: http://nobelprize.org/. Accessed on February 12, 2010.

Periodic Table of Elements. An informative online periodic table of the elements maintained by the Chemistry Division of the Department of Energy's Los Alamos National Laboratory (LANL). Available online. URL: http://periodic.lanl.gov/. Accessed on February 12, 2010.

PhET Interactive Simulations is an ongoing effort by the University of Colorado at Boulder (under National Science Foundation sponsorship) to provide a comprehensive collection of simulations to enhance science learning. The major science categories include physics, chemistry, Earth science, and biology. Available online. URL: http://phet.colorado.edu/index.php. Accessed on February 12, 2010.

ScienceNews is the online version of the magazine of the Society for Science and the Public. Provides insights into the latest scientific achievements and discoveries. Especially useful are the categories Atom and Cosmos, Environment, Matter and Energy, Molecules, and Science and Society.

Available online. URL: http://www.sciencenews.org/. Accessed on February 12, 2010.

The Society on Social Implications of Technology (SSIT) of the Institute of Electrical and Electronics Engineers (IEEE) deals with such issues as the environmental, health, and safety implications of technology; engineering ethics; and the social issues related to telecommunications, information technology, and energy. Available online. URL: http://www.ieeessit.org/. Accessed on February 12, 2010.

Spitzer Space Telescope (SST) is a space-based NASA astronomical observatory that observes the universe in the infrared portion of the electromagnetic spectrum. This Web site contains contemporary information and educational materials about astronomy, astrophysics, and cosmology, including the infrared universe, star and planet formation, and infrared radiation. Available online. URL: http://www.spitzer.caltech.edu/. Accessed on February 12, 2010.

Thomas Jefferson National Accelerator Facility (Jefferson Lab) is a U.S. Department of Energy–sponsored laboratory that conducts basic research on the atomic nucleus at the quark level. The Web site includes basic information about the periodic table, particle physics, and quarks. Available online. URL: http://www.jlab.org/. Accessed on February 12, 2010.

United States Geological Survey (USGS) is the agency within the U.S. Department of the Interior that serves the nation by providing reliable scientific information needed to describe and understand Earth, minimize the loss of life and property from natural disasters, and manage water, biological, energy, and mineral resources. The USGS Web site is rich in science information, including the atmosphere and climate, Earth characteristics, ecology and environment, natural hazards, natural resources, oceans and coastlines, environmental issues, geologic processes, hydrologic processes, and water resources. Available online. URL: http://www.usgs.gov/. Accessed on February 12, 2010.

Index

Italic page numbers indicate illustrations.